THE TOTAL TOMATO

THE TOTAL

TOMATO

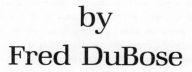

**AMERICA'S BACKYARD EXPERTS REVEAL THE
PLEASURES OF GROWING TOMATOES AT HOME**

by

Fred DuBose

with illustrations by Casey Moore

HARPER COLOPHON BOOKS
HARPER & ROW, PUBLISHERS
NEW YORK, CAMBRIDGE, PHILADELPHIA, SAN FRANCISCO
LONDON, MEXICO CITY, SÃO PAULO, SINGAPORE, SYDNEY

FIRST EDITION

Designer: Barbara DuPree Knowles

Library of Congress Cataloging in Publication Data
DuBose, Fred.
    The total tomato.

    (Harper colophon books)
    Includes index.
    1. Tomatoes.   I. Title.
SB349.D8   1985   635'.642   83-48342
ISBN 0-06-091105-0 (pbk.)

85 86 87 88 89 10 9 8 7 6 5 4 3 2 1

For Dee and Fred

CONTENTS

INTRODUCTION

Finding a decent tomato these days is no easy task. At the very least we have to grow it ourselves or rely on the charity of a garden-tending neighbor. Even home-grown tomatoes may disappoint us, lacking the robust, acid-and-sweet flavor that we wistfully remember as the way tomatoes used to taste; too often they are bland or tough-skinned or hard enough to bounce off a kitchen wall without splitting.

Are we remembering the tomato as something it never was? Hardly. The problem is that tomatoes have been "improved" to within an inch of their lives. Modern varieties grow better, look better and produce bountiful yields in our gardens, but their taste has usually been sacrificed to the needs of commercial growing and marketing. In the 1980s the distinction between home garden varieties and commercial varieties has blurred in seed catalogues and at nurseries, so that we gardeners may innocently be growing tomatoes designed to withstand the abuses of trucking from California to the Midwest when, in fact, we need carry them only from the backyard to the kitchen. Such tomatoes may do exactly what their breeders intended them to do—travel well and fit neatly into styrofoam trays in the supermarket—but rarely do they taste as good as they should.

Yet we should not despair. Those elusive real tomatoes still can be found, in the form of several old-fashioned varieties and many of the new hybrids. They are the tomatoes that were developed with only the home gardener in mind, without the extraneous attributes necessary for large-scale growing and shipping.

This book attempts to clear up the confusion: to distinguish between home garden and commercial tomatoes, between modern hybrids and the older standard types, and to instruct the reader on how to grow them all. Varieties—from Better Boy to Early Girl, from Giant Tree to Tiny Tim—are described in detail and are rated as excellent, good, or fair. And the critics are not professional horticulturists or plant breeders, but typical backyard gardeners who grow tomatoes summer after summer. Thousands of them were asked to evaluate varieties for the purposes of this book, and the results are found in Part 2. In Part 3, many of the same gardeners share their own inventive techniques for getting the best crops. Some of the methods are unusual, even unortho-dox—but all of them work.

The savvy gardeners who care about the varieties they grow, and who are willing to devote a little time to their task, are the ones who will end up with perfect tomatoes. This book is intended to aid them, and to serve as an introduction to the infinite variety of an acidulous little fruit that Americans have come to think of as indispensable.

ACKNOWLEDGMENTS

The author is grateful to each of the 2,000 home gardeners—many of them members of the Men's Garden Club of America—who completed and returned his three-page tomato grower's questionnaire, and to the 350-odd who were interviewed in person or by phone. Special thanks are owed the several plant breeders, garden writers, plant pathologists, entomologists, extension agents, seedsmen, and hobbyists who cheerfully tolerated the author's endless questioning: Randy Drinkard, Joseph Fedrowitz, Derek Fell, J. Danny Gay, John "Mr. Tomato" Gorman, Walter H. Greenleaf, Tom Harding, Frank Hepburn, Victor N. Lambeth, Helen May, Joe McFerran, Howard B. Peto, Leonard Pike, Gene Porter, Charles M. Rick, A. Leon Stacey, M. Allen Stevens, William Taylor, Ted Telsch, Paul Thomas, Dave Thompson, Mark Tomes, Theodore Torrey, James Utzinger, R. B. Volin, Elizabeth Whittle, Jim Wilson, Rob Williams, Roy Wyatt, and the organizers of the Reynoldsburg Tomato Festival, Inc.

Thanks, too, to Stanley Williams for combing the libraries in search of tomato lore, and to Susan Clifton for tabulating a monumental stack of questionnaires.

THE TOTAL TOMATO

PART ONE

HISTORY

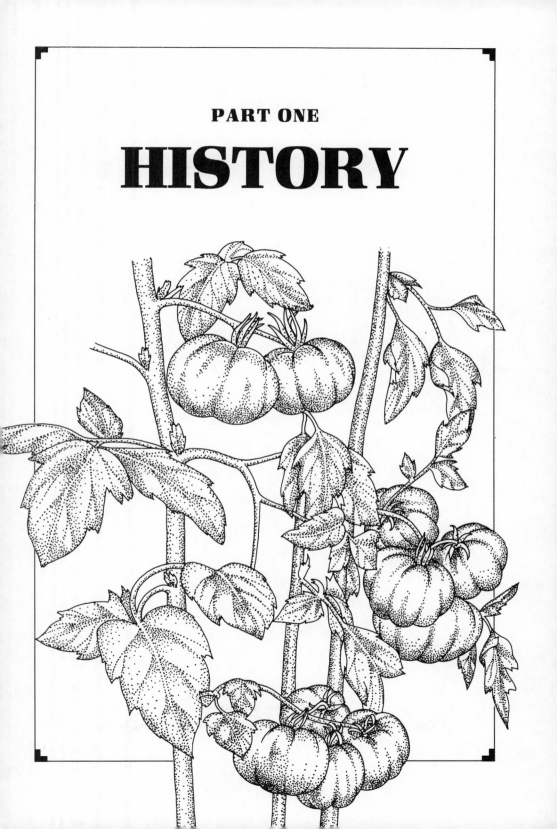

The common tomato seems an odd focus for controversy. Yet for much of its history it has stood accused of one thing or another. Europeans on both sides of the Atlantic shunned it for centuries because they thought it unpalatable, even poisonous. And now, hard and pale on supermarket shelves, it is shunned again—no longer thought of as poisonous, to be sure, but pointed to as modern agriculture's ultimate insult to nature.

The sad decline of store-bought tomatoes has led to new appreciation for the home-grown product—tomatoes that soak up the sun and ripen slowly on the vine in one's own garden. A good tomato, succulent and red, has few culinary rivals. But the normally savory fruit has assumed an almost unrecognizable form in its commercial incarnation.

What went wrong?

THE GOLDEN APPLE

The wild tomato, the parent of the cultivated forms we know today, originated in the northern arm of the Andes that encompasses parts of Peru, Ecuador, Bolivia, Chile, and Colombia. The plants still are found there, sprawling weedlike through tall grass, and would be called cherry tomatoes by a modern gardener who came upon them: fruits are bright red and no more than an inch in diameter, as their scientific varietal name, *cerasiforme* (cherrylike), describes.

The Incas and earlier Andean cultures found little use for the weed. While they may have plucked the ripe fruits in passing, they did not think them worth cultivating—or so it appears. There is no word in South American Indian languages for the tomato, nor is it represented in drawings, as are many common

3

foods, on artifacts and pottery that have been excavated from archaeological sites.

Over centuries the tomato spread northward into Central America and Mexico, its almost indestructible seeds carried by birds and animals. There, more than two thousand miles from its native region, it was domesticated by pre-Mayan Indians. By selecting and planting the seeds of larger and better-tasting fruits, the Indians brought forth the large-fruited tomato that Cortes and his soldiers discovered during the Spanish conquest of the Aztecs (1519–1521). The Aztecs presumably prized a yellow-fruited mutant and cultivated it to become the most common variety, since the first tomatoes taken to Europe were described in early herbals as yellow or golden. The tomato's first common European name translated as "golden apple," and the Italians still call it that: *pomodoro*.

INJURIOUS AND HARMFUL?

Spaniards apparently accepted the alien fruit at first, and introduced it to Italy when Naples came under Spanish rule in 1522. The southern Europeans prepared it as they did another recent import from the New World, the eggplant. Yet they soon became wary. Botanists identified the plant as a member of the family *Solanaceae*, whose members include belladonna, black henbane, and deadly nightshade; and people began to suspect the tomato might be as poisonous as its dreaded cousins.

The first written record of the tomato appeared in the herbal of Pier Andrea Mattioli, an Italian, in 1544: "Another species of mandrake has been brought to Italy in our time, flattened like the melerose [a kind of apple] and segmented, green at first and when ripe of a golden color, which is eaten in the same manner as the eggplant—fried in oil with salt and pepper." Although Mattioli did not state firmly that the tomato was poisonous, and even described how it was eaten, a warning was implicit in its identification as a species of mandrake. (People regarded mandrake, a plant native to the Mediterranean region, with great fear; its poisonous, twisted roots were said to resemble the legs of a human, and it was commonly believed to shriek when pulled from the ground.) A second edition of Mattioli's herbal appeared ten years later with the same description, and noted

THE WOLF PEACH

Botanists assigned the Latin name *Lycopersicon* (wolf peach) to the tomato in the sixteenth century, soon after its introduction to Europe—perhaps because they thought the fruit looked as inviting as the peach, yet was unfit for human consumption. When it was identified as a member of the *Solanaceae* (nightshade) family of plants, its scientific name became *Solanum lycopersicum*. The modern scientific name is *Lycopersicon esculentum*, which translates from the Latin as "edible wolf peach."

The common English appellation comes, by way of a few detours, from the seventeenth century Spanish *tomate*, which was taken in turn from the Aztecs' ancient word for the fruit, *tomatl*. In the 1700s the final vowel changed, and the name became *tomato*, probably in phonetic mimicry of the botanically related potato. For some reason, the form *tomata* became common in the 1800s. Dickens referred to the fruits as such in *The Pickwick Papers*, and Thomas Jefferson, who in his adventurous way grew and ate them long before they were accepted by his countrymen, called the fruits *tomatas* in his diaries. The wandering final "o" returned to the word late in the nineteenth century, and *tomato* settled once and for all into common usage.

the Italian name for the fruit, *pomodoro* (golden apple), and its Latin equivalent, *mala aurea*. It also mentioned that there was a red variety of the fruit.

An illustrated Italian herbal that appeared later in the century revealed more than an implied distrust. Pietro Antonio Michiel, in the label for a drawing of the tomato, wrote, "If I should eat of this fruit, cut in slices in a pan with butter and oil, it would be injurious and harmful to me." It can be presumed that the tomato had been used as food, since the writer gave an appetizing recipe for its preparation, but that it still was under suspicion. The suspicion was, of course, without substance. The leaves and stems of the plant are mildly poisonous, but the toxic alkaloids degrade to inert compounds in the fruit. The innocent tomato seems to have been guilty simply by association with its botanical relatives, since it is certain that no one could have become ill from eating it. Perhaps there were cases of hapless diners consuming the leaves (a mistake common with the potato after its introduction) and later complaining of stomachaches, or perhaps an occasional allergic reaction gave the fruit a bad name.

In 1574 came a report that was to be repeated in herbals for the next three centuries, confirming, at least in the minds of those who could read, the notion that the tomato was dangerous. It appeared in the herbal of Rembert Dodoens of Antwerp: "The [tomatoes] are eaten by some prepared and cooked with pepper, salt and oil. They offer the body very little nourishment and that unwholesome." Dodoens's conclusion that the tomato was un- wholesome was accepted without question by herbalists for years to come. His description of the fruit appeared time and again in other books, sometimes slightly paraphrased, sometimes verbatim. (Herbalists of the time seemed little concerned with originality, and even less with conspicuous plagiarism.) John Gerard, the first Englishman to write of the tomato, expanded creatively on Dodoens's words: "In Spaine and those hot regions they used to eat the [tomatoes] prepared and boil'd with pepper, salt and oil; but they yield very little nourishment to the body and the same naught and corrupt."

Later herbalists, if they did not borrow so shamelessly from Dodoens, at least served to heighten the anxiety. Hermann Boerhaave stated in 1727 that the tomato's "seeds when taken upset the stomach and cause faintness and a sort of apoplexy." Such ridiculous claims eventually died down, but as late as 1856 a French herbalist still expressed some misgivings: "It is found that the use of these fruits, especially when overripe, is sometimes followed by colic or diarrhoea."

These queer allegations may have merely reflected the people's distrust of the tomato. There was an aura of something forbidden, of bad luck, about the plant. Aphrodisiacal properties even were attributed to it, perhaps on the puritanical assumption that anything that looked so alluring—so smooth, round, and scarlet— had to have, like Eve's apple, a wicked side.

The tomato's reputation as an aphrodisiac has probably been

exaggerated in literature. From the beginning, one of the names used to describe the fruit in Italy was *pomum amoris* (apple of love), and the earliest common English name for the fruit was "love apple." But the first to call it that probably did so for the beautiful flame-red color of the fruit rather than for any amatory qualities. (The red variety had become the most common by the end of the sixteenth century.) One writer has come up with the imaginative, if dubious, theory that the French were responsible for assigning the name "love apple" to the plant through a misunderstanding: an early Italian name, *pomi dei Moro* (Moor's apples, in a time when Spaniards were sometimes called Moors) sounded to the notoriously romantic French like *pommes d'amour.*

"ATROCIOUS IN FLAVOUR"

Early herbals also suggest that the taste and smell of the tomato were partly responsible for its rejection. Perhaps to Europeans the taste was simply unfamiliar; or it could have called nightshade to mind, an association that meant poison. Or, conceivably, the early cultivars really did have a disagreeable taste. Written records of the period contradict one another. In Germany, Gesner described the tomato in 1561 as "odorless, not unpleasant, not harmful in food." Yet, in 1583, the Italian Caesalpinus wrote that tomatoes were better grown as ornamentals than as food, since "they have a certain musky odor, particularly the red ones." It is ironic that the German, not the Italian, declared tomatoes suitable to eat. Italy, after all, was the country that eventually brought the fruits to gastronomic glory.

Even the French, normally adventurous in the kitchen, seemed to dislike tomatoes at first, and in a slightly hysterical way. A French herbal from 1600 stated that "this plant is more pleasant to the sight than either to the taste or smell, because the fruit being eaten provoketh loathing and vomiting."

While the tomato was neglected as food, it was flourishing in the garden. Gardeners prized it as a fast-growing ornamental to cover arbors and outhouses, and placed specimens in pots so the fruits could be admired. The English grew it as an ornamental for two hundred years before daring to eat it.

There was also a practical use for the plant. The juice of its leaves and stems was used as a medicine, first in the sixteenth century, to treat skin diseases. The medicinal value was no old wives' tale: scientists discovered in the 1940s that an ingredient in the leaves, named *tomatine*, did indeed stop the growth of fungi and parasitic yeasts that cause ringworm and athlete's foot. And users of the expensive tomato soap found today in speciality stores swear by its effectiveness for soothing rashes and inflammations.

Other medical claims made for the tomato, however, were questionable. A Flemish herbal in 1635 described tomato juice as a remedy for an upset stomach. In the seventeenth century the tomato was said to cure glaucoma, to moderate fever, and to yield an oil that "rubbed on the temples and body induces sleep." A British herbal claimed the tomato "represses vapors in women." By 1838 an American magazine, *The Cultivator*, said that tomatoes "are made the basis of a medicine which, if we are to credit the declarations of the vendors, is an infallible cure for most all sorts of diseases which man is heir to." Vendors claiming infallible cures call snake-oil salesmen to mind, but a statement by the same magazine four years later gave the medical claims more respectability: "If the opinions of numerous M.D.s of great celebrity are to be allowed of any weight, there are few things more conducive to health than a liberal use of tomatoes." The tomato had, in the course of three centuries, changed from feared poison to marvelous cure-all.

"EVERY BODY CULTIVATES THE TOMATO"

In North America, as in Europe, tomatoes had been grown as ornamentals for years, but few people chose to serve them at the table. French settlers in New Orleans and Quebec used them, mainly for catsup, in the late 1700s, and Thomas Jefferson, who made a hobby of growing exotic foods, planted them at Monticello as early as 1782. Records from the Philadelphia market show that tomatoes were sold there in 1784, but that they, along with okra and artichokes, "were first demanded by the French immigrants and there was little sale for them to others."

Tomatoes still were rare in markets in the early 1800s—partly because their perishability made shipping difficult, but mainly because they still were not accepted by most people. Yet we can judge from cookbooks of the time that they were grown in the kitchen gardens of those who, like Jefferson, had a certain degree of culinary sophistication. Tomatoes probably occupied the same place in the culinary spectrum that salsify or celeriac, for example, occupy today: they were appreciated by some, but hardly were common in the diet of the average person. Among the uneducated and superstitious, they still were thought to be highly dangerous.

One event—the consumption of a basketful of tomatoes in public by one Robert Gibbon Johnson of Salem, New Jersey—is sometimes cited by food historians as the turning point in America's acceptance of the tomato. It is farfetched to suppose that three centuries of suspicion could be dispelled in one afternoon, by one man (and in a remote New Jersey village, at that), but the story has the kind of irresistible drama that impels us to believe it changed the course of history. Hence Johnson's celebrated, if dubious, place in tomato lore.

Robert Gibbon Johnson was wealthy, respected, and eccentric. And, for a reason that may never be known, he took it upon himself to popularize the tomato. He encouraged local farmers to plant the crop, but without success. The townspeople of Salem, the story goes, were convinced tomatoes were poisonous, and that no one could eat one and survive. One day in 1820 (some accounts say 1830), Johnson set out to prove them wrong: he announced he would eat not one, but several, tomatoes in front of anyone who wished to watch. At high noon, Johnson ascended the courthouse steps and consumed a basketful of fresh tomatoes from his garden in full view of the crowd that had gathered. It is said that when the defiant gentleman suffered no ill effects, the tomato was immediately accepted, thereafter gracing not only the tables of the Salem townspeople, but also, presumably, those of the rest of America. Some historians have embellished the event to include marching bands and thousands of spectators, many of whom fainted at the sight of a tomato being bitten into.

Johnson's brave act did coincide with a growing acceptance of the tomato. But Americans accepted the fruit for less dramatic reasons. New varieties tasted better, and improved transportation allowed them to be carried longer distances to market. And the poison myth, with no basis in fact, gradually just faded away.

By the 1840s the tomato actually had become fashionable. In 1842, *The Cultivator* magazine stated, with missionary zeal, "Every body cultivates the tomato and every one who has not deliberately made up his mind to be ranked among the nobodies has learned to eat it. There is a great deal of fashion in this, it must be confessed, but it is not often that fashion is active in forwarding so good a work." There follows a paean to the healthfulness of the fruit.

Although in vogue, the tomato still was used for the most part in pickles, preserves, and catsup, or to add tartness to sauces. Eating it plain, as we do today, seems to have been an acquired taste. Cookbooks of the time advised that tomatoes should be cooked for at least three hours to "lose their raw taste." A farmers' magazine stated in 1842 that "there are but few who relish the tomato at the first taste, and few who are not extremely fond of it when properly cooked and they become accustomed to it."

THE TOMATO IS PERFECTED

By the 1860s, commercial seedsmen were vying to come up with ever better varieties, so that people no longer had to learn to eat tomatoes, nor had to cook them to enjoy them. They improved the fruit's flavor and streamlined its form, since tomatoes of the time still were ribbed or fluted and often grew in grotesque shapes. A few of the varieties developed in the first flush of American tomato breeding still are available today. Red Pear dates from before 1850, Crimson Cushion (which survives today as Beefsteak) was developed around 1855, and Yellow Plum was introduced in 1865. Most, however, have long since disappeared. Among the most popular varieties was an 1870 introduction called Paragon, bred by seedsman Alexander W. Livingston of Reynoldsburg, Ohio, and promoted as "the first perfectly smooth, deep red tomato ever offered to the American people." (Today, signs on the outskirts of Reynoldsburg proclaim it the "Birthplace of the Edible Tomato"; its citizens, in the delightful tradition of small towns, celebrate Livingston's contribution with an annual Tomato Festival complete with carnival rides, a grand parade, and a Tomato Queen.)

Seedsmen improved tomatoes in much the same way the Indians of Central America transformed the original wild cherry tomato into the larger fruit the Spaniards discovered in Mexico: through the horticultural process of selection. They selected and propagated only the seeds of superior specimens, gradually breeding out such undesirable qualities as sourness, small size, or irregular shape. Tomatoes were also improved through hybridizing—crossbreeding two varieties and selecting the progeny that had the best characteristics of both.

When a lucky "sport" turned up—a rogue plant that, through genetic mutation, produced fruits with strikingly different characteristics from others of its variety—seedsmen propagated it year after year until the seeds of its progeny were certain to "come true." In this way, a new variety—a true-breeding strain—was born.

Thus were tomatoes perfected, the product not only of nature, but also of horticulture. By the end of the nineteenth century, the consummate tomato was smooth, richly succulent, and flavored with a fine balance of sweetness and acidity.

Increasing public demand for these new-and-improved versions led to their use as a commercial crop in the 1870s. Farmers harvested them when they were almost ripe and carted them to market in nearby towns. Some growers set up commercial canneries, so that Americans could enjoy their new delicacy year-round. The first to do so was Harrison W. Crosby of Easton, Pennsylvania, in 1848. More successful in their venture, in 1869, were Joseph Campbell and Abraham Anderson of Camden, New Jersey, whose cannery became the Campbell Soup Company.

At the turn of the century, and for the next two decades, gardeners were presented with tomato varieties of sundry shapes, sizes, and colors. The names of many of them still are familiar: Ponderosa, Bonny Best, Dwarf Champion, Earliana, Break o' Day, Oxheart. And, as commercial production grew in scale in the 1920s, varieties with another characteristic, firmness, were developed to better withstand the abuses of shipping—a portent of things to come. By the 1940s, the tomato had not only arrived, but had become big business, and North Americans would have been hard pressed to imagine a time when tomatoes were not as easily available as soap powder or soda pop.

FRUIT . . . OR VEGETABLE?

Botanically, the tomato is a fruit—a structure that develops from the ovary of the plant after fertilization has occurred. More precisely, it is a berry, since it is fleshy and contains several seeds. (Such precision may serve only to confuse the issue, however, since the berry family technically includes such unlikely members as grapes, oranges, lemons, squashes, and cucumbers, as well as tomatoes; and many of the fruits we regard as berries, such as blackberries and raspberries, really are not berries at all, but simply clusters of fruits.)

Ignoring botany and categorizing the tomato by its use—as a side dish, in salads, or as an ingredient in cooking—it is a vegetable. The law says so. The U.S. Supreme Court ruled in 1893 that because the tomato was used as a vegetable—that is, was eaten with the main part of the meal instead of with dessert—it should be legally classified as such, and therefore was subject to tariffs when imported from abroad. (All imported vegetables were taxed at the time, but fruits were not.) The ruling effectively curtailed the importation of tomatoes from Mexico and Cuba and resulted in a bonanza for U.S. commercial growers.

THE DECLINE

What, then, went wrong? Why did the store-bought tomato, which had been getting steadily better for almost a century, suddenly start getting worse? Why did it turn hard and mealy, and lose its taste? That question is asked by virtually anyone who bites into the commercial product, and the answer is simple: the tomato became too popular for its own good. Like anything worth having in the postwar consumer wonderland that America became, it was expected to be available anywhere, anytime, in keeping with the newfound notion that the food supply was not beholden to the seasons, but merely to the whims of consumer demand. Tomatoes might have been a summer crop, but in the fluorescent-lit world of the supermarket, the seasons never changed.

Americans expected to buy tomatoes as easily in January as in July, so agriculture found a way to supply them. But there was a problem. The juicy, vine-ripened tomatoes that had created such a demand in the first place were impossible to provide out of season except in the frost-free southernmost reaches of the country. By the time they were picked, shipped even a few hundred miles, and placed on a grocer's shelf to await the buyer's pleasure, they would have begun to leak and rot, tomatoes being among the most perishable of common foods. So plant breeders stepped up their improvement of the tomato. Unfortunately, improvement no longer meant "better tasting"; it meant, simply, "more marketable."

Commercial tomatoes, grown in sunny Florida and California, had to be made fit for marketing in the Midwest and North. This meant they had to be harder, so they could withstand mass harvesting and shipping and display without being mashed or split. It meant that a tomato plant should ideally set almost all of its fruits at the same time, so that pickers could harvest them in the space of a few days, and keep labor costs down. And, most regrettable of all, it meant picking tomatoes green and letting them ripen en route to market instead of on the vine.

The result was tomatoes that were tasteless and dry. But their decline was so gradual that consumers seemed unaware, or perhaps unconcerned, that the modern product bore little resemblance to the original. Demand continued unabated. The public seemed altogether willing to buy tomatoes as long as they were round, smooth, and packed in sanitary plastic trays. In an age of synthetic convenience foods and TV dinners, such uncritical acceptance was perfectly in character. By the time consumers woke up and took stock of the tomatoes that were being sold them, the modern tomato had degenerated to a state even sorrier than that of its maligned ancestor.

PRODUCTION IN THE MODERN AGE

Plant breeders, continually dispensing new tomato varieties from private seed companies, state universities, and agricultural experiment stations, are touchy about criticism. They say consumers do not understand the economic realities of the large-scale commercial production that makes it possible to have tomatoes year-round at a reasonable price. And besides, they say, it is not so much the varieties themselves that are at fault as the handling—the early picking and shipping by growers and packers—that makes supermarket tomatoes less than they should be.

Breeders point with pride to the advances that have made it possible to genetically engineer a tomato down to the last detail. More and more F_1 (first generation) hybrids have been developed over the past thirty years, and the discovery of various mutant genes has led to the introduction of both hybrid and standard varieties that are tailor-made for special purposes. A mutant gene, called by scientists sp for self-pruning, revolutionized commerical production by making it possible to breed tomato plants that stopped growing after they reached a certain size. (Such varieties are said to be determinate, as opposed to the ordinary indeterminate varieties that keep growing until killed by frost.) These plants, left unstaked, do not sprawl into irrigation ditches, where fruits would rot; and they tend to have a more concentrated fruit set, so that most fruits ripen at the same time and can be harvested in fewer pickings. Another gene, u for uniform ripening, programs tomatoes to ripen without green shoulders, so that the whole fruit can be eaten. Breeding stock is now so varied that horticultural scientists can assign any number of characteristics to a given variety: early ripening, fruit-setting at extremes of temperature, firmness, disease resistance, concentrated fruit set, crack resistance, consistent fruit size, and heavy, protective foliage. Various combinations of these and other qualities make it possible to design tomatoes to suit every need of commercial growers, from fitness for transport to lower labor costs.

That, consumers argue, is the problem. Tomatoes are designed for the people who grow them, not for the people who eat them. At the center of the controversy is the "fresh-market"

GOOD FOR YOU?

In its concentration of ten essential vitamins and minerals, the tomato ranks only 16th among common fruits and vegetables. More nutritious, in descending order, are broccoli, spinach, Brussels sprouts, lima beans, peas, asparagus, artichokes, cauliflower, sweet potatoes, carrots, corn, potatoes, and cabbage. But in its total contribution of those ten vitamins and minerals to the diet, the tomato ranks first, simply because it is eaten so much more frequently than the other vegetables.

If a typical, medium-sized tomato carried the nutrition information that the U.S. government requires for packaged foods, its label would read as follows:

SERVING SIZE: Weight, 4 $^3/_4$ oz.; diameter, 2 $^3/_5$ in.

SERVINGS PER TOMATO: 1

CALORIES: 27

PROTEIN: 1.4 g

CARBOHYDRATE: 5.8 g

FAT: 0.2 g

SODIUM: 4 mg

		U.S. Recommended Daily Allowances (U.S. RDA)
VITAMIN A	1110 i.u.	20%
VITAMIN C	28 mg	45%*
THIAMINE	0.07 mg	4%
RIBOFLAVIN	0.05 mg	2%
NIACIN	0.9 mg	4%
CALCIUM	16 mg	**
IRON	0.6 mg	4%
POTASSIUM	300 mg	***

* Year-round average; tomatoes grown from June to October have slightly more Vitamin C; those grown from November to May have slightly less.

** Contains less than 2% of the U.S. RDA of this nutrient.

*** No U.S. RDA has been established for this nutrient.

tomato, known to the public as the supermarket tomato.

Most people understand that it is impractical, in a self-service economy, to sell ripe, soft, fragile tomatoes of the old-fashioned kind. But they question whether it is really necessary for super-market tomatoes to be so awful. Large-scale commercial growers provide the answer. In the multimillion-dollar tomato industry, they say, the main concern is to get tomatoes produced in the quickest, most cost-efficient, profit-making way. Nowadays, many fresh-market tomatoes are still a labor intensive crop, handplanted and handpicked by laborers. But there is an inevitable move toward mechanical harvesting, in which huge mobile machines— traveling factories of a sort—lumber through tomato fields picking fruits, sorting them by size and moving them along conveyor belts into bins. As a result, more fresh-market varieties are being bred with characteristics that suit the machine: tough skin that won't split when the tomatoes are banged about in the harvester, and even "jointlessness," which means tomatoes can be detached from the plant free of their stems. Other kinds of tomatoes, the processing types intended for use in commercial sauces, catsups, and soups, are being bred with a slightly square shape to keep the fruits from rolling off slanted conveyor belts and to prevent splitting and leakage when they are trucked en masse to the processing factory. These characteristics are incorporated on top of the usual commercial qualities of firmness, high yield, disease resistance, uniform ripening, and any number of others. Once a tomato has been designed to do so many things, there is little room left for flavor. In the production of a cost-efficient tomato, taste is a secondary consideration, at best. The scientific tomato begins to seem less the product of sun and soil than the mass-produced yield of a plastics factory.

Whether harvested by hand or machine, most fresh-market tomatoes are picked at the "mature-green" stage—when the fruits are still completely green, yet the juice in the seed cavities has begun to jell. They are then placed in gas-filled rooms to speed ripening.

The gassing of tomatoes, with its suggestion of chemicals and artificiality, has become a cause célèbre among critics. In truth, it is not as bad as it sounds. The gas is neither dangerous nor artificial, but is merely the natural ripening agent that fruits produce themselves: ethylene. To ripen mature-green tomatoes, ethylene is pumped into the gassing rooms, where tomatoes bask in it for at least a day, by which time a faint blush of pink appears at the blossom end of the fruits. Once spurred by the gas, the ripening process will continue on its own in transport trucks, on supermarket shelves, and on kitchen windowsills. Harmless though it may be, gas-speeded ripening is a much different proposition from letting tomatoes ripen on the vine. The production of ethylene is only one part of the natural ripening process. The full transformation from green to ripe fruit can happen only on the plant, when the slow manufacture of sugars, acids, and aromatic oils combine to give the tomato an exquisite flavor that cannot be duplicated outside of nature.

Taste, after all, is what we eat tomatoes for. In the effort to provide tomatoes year-round on a mass scale, breeders and growers have managed to lose the very thing that made the tomato so sought after in the first place: its rich tang and succulence. They have succeeded in giving us a mere image of the perfect tomato—round and smooth and blemish-free, but, when bitten into, insipid.

RECLAIMING THE TOMATO

Someday, science will find a way to breed taste back into tomatoes. Research has been under way since the late 1970s to identify the volatile compounds and the sugar and acid concentrations that give tomatoes their flavor, so that breeders can select for taste as well as for the qualities that make commercial growing feasible. (Consumers may be forgiven for asking what took them so long.) In the meantime, North Americans have little choice but to grow their own if they want to reclaim the pleasure of eating real, vine-ripened tomatoes. Forty-eight percent of households in the U.S. have some sort of vegetable garden, and the percentage rises every year. Many of those gardens may be no more than a couple of tomato plants in the corner of a flower bed or in pots on the patio, placed there by people who otherwise would never dream of growing their own food, yet are determined to eat real tomatoes instead of the pretenders from the supermarket.

Real tomatoes, after all, have never really disappeared. They are available to us today in dozens of forms, from nineteenth century heirloom varieties to some of the modern hybrids. But now they compete with the countless commercial tomatoes that are sold for growing at home, and it is harder than ever for the home gardener to distinguish those meant for large-scale growing from those meant for backyards. Gardeners who make an effort to learn the difference, however, are the ones who stand to reap a real reward: the perfect tomato they have almost given up hoping for.

PART TWO
VARIETIES

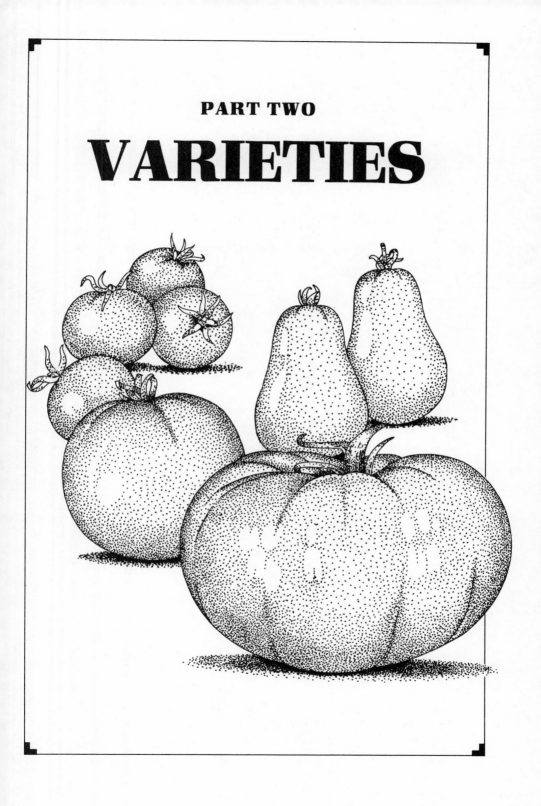

To most people, a tomato is a tomato. If any differences are acknowledged, they are usually the most obvious ones. Tomatoes, people say, are either large or cherry-sized, round or plum-shaped, and either bought at the grocery store, in which case they are awful, or grown in a friend's garden, in which case they are wonderful.

Gardeners who grow their own will probably make a few more distinctions, knowing when their tomatoes will ripen on the vine—early, late, or midseason—and whether they are the old-fashioned open-pollinated types or one of the newer hybrids.

For the gardener or the connoisseur who wants to know more, however, there is a good deal more to consider. There are real, noticeable differences between one variety of tomato and another; and hundreds upon hundreds of varieties are sold, no two of them alike. Natural genetic mutations and scientific crossbreeding have brought forth amazing transformations of the oversized berry the Aztecs called *tomatl* more than four centuries ago. Some tomatoes are not even red: many are rose pink, some are brilliant yellow, and a few have hardly any color at all. Some are tailor-made for Tucson, others grow well in the humidity of New Orleans or the cool summers of Newfoundland. Some tomato plants are neat little bushes that support themselves, while others run rampant through the garden or outgrow the stakes to which they are tied.

From a gastronomic standpoint, a tomato is most decidedly not just a tomato, especially at a time when the North American palate is becoming increasingly sophisticated. This is an age when the subtle distinctions between varieties of foodstuffs, from cheeses to wines to vegetables, are recognized and appreciated. Tomatoes should be enjoyed in the same way. The difference

21

between the taste of a Big Boy and a Tiny Tim may not be as great, by any stretch of the imagination, as that between Gouda and provolone, or between Burgundy and Moselle; but there is indeed a difference, even if it is highly unlikely that a precise vocabulary will ever exist for tomato taste as it does for wine. Some varieties are sharp, even biting in flavor; others are almost sugary enough to qualify as desserts; some are subtly aromatic, while others can only be described as bland. The tomato's taste, a combination of the sugars, acids, and volatile oils that form slowly in the fruit as it ripens on the vine, is as varied as its color, its size, and its growth habit.

Casual gardeners may not care about all this diversity. They assume that any home-grown tomato is going to be better than its supermarket counterpart, and, as far as cultivation is concerned, don't all tomato plants just "grow themselves"? It is true that even the most haphazard gardener can choose any one of the better-known varieties, leave it alone, and probably come up with some kind of crop, barring drought and disease. And it is not exactly essential for gardeners to know much about the differences between tomatoes, especially when they can depend on the local nursery to sell only the varieties that do well in the local climate.

Serious gardeners, however, should know that there may well be better choices than the varieties sold at the nursery or pictured most prettily in the mail-order seed catalogue. Even nurseries are hard pressed to keep up with the constant introduction of new tomatoes, and may sell the same old varieties out of habit, or to supply gardeners who demand certain tomatoes simply because they are used to them. It is often said that the average home gardener is twenty years behind the times when it comes to choosing varieties. By the same token, nurseries may be behind the times in the varieties they offer.

Those gardeners who take the time to acquaint themselves with the huge range of varieties available, and to choose carefully, with their own particular set of garden conditions in mind, can reap the backyard gardener's ultimate prize: the perfect tomato. Moreover, those who are determined to come up with a tomato that is unusual, oversized, or novel can gain an edge over the vegetable-growing neighbors with whom they may have a friendly competition. And all gardeners, simply by knowing something of the history of the varieties they are growing, and the reasons for their development, can add another measure of interest to the pleasure of gardening.

COMMERCIAL VERSUS HOME GARDEN VARIETIES

Before a gardener chooses a variety—whether ordering seeds from a catalogue, buying transplants at a nursery or selecting a seed packet at the hardware store—he should be aware that there are two broad categories of tomatoes: commercial and home garden.

Commercial tomatoes were bred to suit the needs of mass producers who plant and harvest on a huge scale. They are perfect specimens for commercial growers, who point to them as supreme examples of the plant breeder's art because they are hard, they are the same size, they withstand any amount of abuse—in harvesting, shipping, and in handling by shoppers at the supermarket—and they keep well on the grocers' shelves. And, most amazing of all, they are always available, even in the dead of winter.

It is obvious to almost everybody, however, that the essential attribute of the tomato—its taste—has been lost in the effort to keep everyone's kitchen well stocked with tomatoes year round. Plant breeders have placed more emphasis on high yields, disease resistance, and other qualities than they have on flavor. And the grower's objective is to get a tomato to the grocer's shelf without spoiling, which deprives the fruit, by necessity, of the very process that gives it taste: vine ripening. Picked green from the vine, the tomato will ripen, or at least turn red; but its taste, and even its nutritive value, never will approach that of its vine-ripened counterpart. The progressive reddening of tomatoes in ethylene-filled storage rooms, transport trucks, and supermarket produce sections is simply no substitute for ripening on the humble little plant.

What of the supermarket tomatoes labeled "vine ripened"? Unfortunately, the claim is an empty one. In the world of large-scale tomato marketing, "vine ripened" means only that the ripening process *started* while the tomatoes were still on the vine. That is, they were picked when the first blush of pink appeared at the fruit's blossom end, instead of being picked wholly green, as most commercial tomatoes are. This somewhat delayed harvest improves them slightly, but not enough to save them from mediocrity.

Gardeners, perhaps encouraged to grow their own tomatoes precisely because of the poor quality of supermarket ones, may assume the varieties available through mail-order catalogues and local nurseries will not have the characteristics that commercial growers seek and home gardeners avoid. Can't they grow any variety bought from a catalogue or nursery and trust it to give them the real tomatoes they want? Perhaps, but not necessarily. In an age of literally thousands of varieties, hundreds of which are available for growing at home, distinctions have blurred between commercial, processing, and home garden types—and many, if not most, tomatoes sold through catalogues and nurseries *are* commercial varieties. Some catalogues list home garden and commercial types separately, but others provide only offhand mentions within variety descriptions, or give no clues at all. (Not that catalogues try to deceive the gardener; many merely either assume knowledge on the consumer's part or do not think the distinction worth worrying about.)

This is not to suggest that all commercial varieties are bad and all home garden ones are good. In many cases, the differences are so minor as to be meaningless. Gardeners who want special characteristics—whether a certain plant size and growth habit, a specific disease resistance or even a particular flavor—may find that a commercial variety suits their needs better than a home garden type. And any commercial variety will fare better when grown at home and picked dead ripe from the vine. It will usually be harder—and in most people's opinion, less tasty— than a home-grown tomato needs to be, but it will be considerably better than its anemic kin in the supermarket.

Nevertheless, to pose a question, why would a gardener who wants a flavorful, juicy tomato which has to be carried only from the backyard to the kitchen, grow a variety bred to be carried from a commercial field to the supermarket? The qualities of commercial varieties—firmness and tough skin, among others— are altogether extraneous in the home garden. Growing a jointless variety that was designed for mechanical picking, or one bred to have uniformly sized fruits that look good in styrofoam trays, just does not make sense.

While commercial tomatoes may or may not suit a gardener's needs, there are, we can be thankful, dozens of varieties bred specifically for home gardens—tomatoes that were never meant to be efficiently harvested, gassed, and shipped, and have no other purpose than to taste good. Some of them are the old standbys developed more than half a century ago. Others are modern hybrids, supertomatoes of large size, high yields, and specific disease resistance, yet without the commercial qualities that are irrelevant in the home garden.

How is the gardener to tell the difference when he chooses varieties to grow? A few clues in seed catalogues or nursery descriptions can help. Commercial types are made obvious by such phrases as "suitable for basket trade," "best variety for green pick shipping," "excellent for the canner trade," and "firm in both the green and red stage." Other indications are less apparent: the designation "pak" in the name of the variety (Basket Pak, Early Pak, and Stokes Pak are examples) suggests the variety was bred for packing and shipping, hardly a recommendation to home gardeners; "jointless" means the tomato can be detached from the vine free of its stem—important for growers who use huge mechanical harvesters, but of no consequence to anyone else; "good replacement for MH-1" (or any other unfamiliar variety with a suspiciously impersonal name) implies bigger marketing potential rather than better tomatoes; "processing" means intended for use by commercial canners or catsup makers.

Certain terms are commonly used in catalogues and on nursery labels to describe the characteristics of both commercial and home garden varieties. Some tomatoes are frequently described as "good slicers," meaning they are meaty enough for a slice to hold together well, and are not so juicy that sandwich bread will become soggy. "Meaty" means the proportion of flesh to seeds and gel is high; these types are sometimes described as "thick-walled," although thick walls may be a hint that the variety was bred to withstand shipping and handling without splitting. Thick walls are also a characteristic of paste tomatoes used for cooking. "Solid" is often a synonym for hard, and is sometimes used interchangeably with "meaty." "Mild" means the tomato *tastes* less acidic than most (even though it may have just as high an acid content). "Smooth" means that the tomato's outer surface is free of ridges or fluting. "Perfectly round" suggests the tomato does not have prominent shoulders. (Gardeners may question why they should care whether a tomato has prominent shoulders. It is because there is less waste in tomatoes without shoulders—a big depression in the center of the stem end means big holes in the last couple of slices of tomato.) "Delicious" and "full-flavored" are often accurate descriptions of the variety in question, but sometimes are meaningless exaggerations on the part of the copywriter.

THE BASIC TYPES

Tomatoes come in several shapes and sizes, from round, bite-sized cherry types to huge, irregularly shaped beefsteaks.

Although there is a specific variety called Beefsteak, the word is most often used generically to refer to very large, late-maturing tomatoes—those that weigh more than a pound each. (Tomatoes sometimes grow up to 2½ pounds, and Guinness says the world record is held by a 6½-pound one grown in 1976.) The name, with its intimations of meatiness and flavor, originated when the Anderson & Campbell Company, the first to can tomatoes, packed one large tomato to a can and sold them under the brand name, "Beefsteak"; on the label was a fanciful picture of two farmers hauling an enormous tomato the size of a horse-drawn buggy. The company eventually went on to greater success as Campbell Soup.

Beefsteaks are what many people mean when they speak of old-fashioned tomatoes. Their taste is sweet and aromatic, with only a hint of acidity. As their name suggests, they are the meatiest of all tomatoes. Older varieties are irregular in shape, and have big cores and prominent shoulders. (Before Anderson & Campbell popularized the term "beefsteak," these types were known as cushion tomatoes, because their big shoulders made them resemble a cushion with a deeply recessed center.) They are the imperfect tomatoes that modern plant breeders have tried to improve.

Cherry tomatoes are the other extreme, measuring from an inch or less in diameter to almost two inches. They are generally sweeter than larger types, are perfectly round and smooth, and have thinner skins. Some of them grow on compact little plants that are ideal for containers, and some will sometimes even produce a crop of tomatoes indoors if placed in a sunny window.

Plum and pear tomatoes, developed more than a century ago in Italy, take their name from their shapes. Plum tomatoes are fairly small, although usually not as small as cherry types, and are slightly elongated. They have firm flesh, with few seeds and little gel, and cook down to a thick paste or sauce. They can also be eaten fresh, of course, though their lack of juice makes them less desirable to most people than the ordinary types. Pear tomatoes are small and have a distinct neck. They are used in the same way as plum-shaped ones, and also for pickling and making preserves.

Much has been written in recent years about so-called square tomatoes, and many home gardeners are curious about them. Square tomatoes are not designed for the home garden, however, but for mechanical harvesting; their shape keeps them from rolling down the slanted conveyor belt of the harvesting machine. They are used for processing by commercial canning companies. Home gardeners who want to grow them as novelties should be aware that the tomatoes are not actually square, but merely slightly elongated and flattened at both ends. They look not so much square as malformed.

COLOR

Tomato varieties differ in the proportions of the various carotene pigments that give them color. Most of them are high in lycopene (the pigment that makes tomatoes red), and have a smaller proportion of betacarotene (the yellow-orange pigment that gives carrots and other orange vegetables their color and is also the source of vitamin A). Red tomatoes range in color from rich red-purple to fiery red-orange.

Not all tomatoes are red, however, and not all depend merely on the amounts of various carotenes for their color. Pink tomatoes (pink when fully ripe, that is—not pink in the manner of unripe supermarket tomatoes) actually have red flesh, but appear pink on the outside because their skins are translucent; the red-fruited flesh shows through as pink. Pinks (with the old beefsteak variety, Ponderosa, as the best-known example) are usually described as milder-tasting than reds. But this mildness has nothing to do with pigment. It is there because most of the pink types are meatier; and the fewer the seeds and the less the gel, the milder the taste of a tomato.

Yellow tomatoes are considered by many gardeners to be freaks—some trick played by plant breeders—out of a stubborn belief that the only real tomatoes are red. On the contrary, the first tomatoes introduced to Europe in the early sixteenth century were yellow, and were known as "golden apples." Another misconception is that yellow tomatoes are lower in acid than red types. Even though laboratory tests show that yellows are just as acidic as reds, the belief persists because yellows *taste* milder. As with pinks, most of them taste milder because they happen to have a higher proportion of flesh to seeds and gel.

Yellow tomatoes are yellow simply because they lack the red carotene called lycopene. Older yellow varieties, such as Yellow Plum, are bright yellow, while many newer ones, such as Jubilee and Sunray, are yellow orange. The orangeness of the newer varieties is not, as some people assume, the result of a higher betacarotene content, but is caused by a mixture of other carotenes instead. In fact, these yellow-orange types have no more beta-carotene than red tomatoes, and therefore are no more nutritious.

In short, color makes no difference to flavor. But psychology and expectations seem to play tricks on the taste buds. This was demonstrated in a test conducted by horticulturists from Purdue University at the Indiana State Fair in the 1960s. More than 3,000 people were asked to compare the taste of an ordinary red tomato with that of an orange-red one (orange enough to look distinctly different from the typical tomato). The red tomato was favored by almost everyone. But when the same people were asked to taste the same tomatoes in a booth fitted with a special light filter to mask the color of the fruits, hardly anyone preferred one tomato over the other. Color, it seems, plays a large part in the psychology of taste.

So-called white tomatoes, such as White Beauty and White Wonder, actually have pale yellow flesh, but look almost white on the outside because their skins are translucent. Many people assume that they are the least acidic of all tomatoes, because they are the palest. But, again, color has nothing to do with acidity. Meatiness does, and whites, like the pinks and yellows, are fleshy and have few seeds.

EARLY, MIDSEASON, AND LATE VARIETIES

Most tomato varieties—the ones called midseason or main crop tomatoes—supply ripe fruits from 65 to 90 days after transplanting to the garden. Those that take as long as 80 days or more are sometimes called late varieties.

Others, however, produce ripe tomatoes more quickly. These early varieties can give the gardener table-ready fruits as soon as 50 days after seedlings have been transplanted to the garden. (Seedlings are usually 3 to 5 inches high when transplanted, and have not yet begun to bloom with the flowers that will give way to fruits. Some keen gardeners will start their seedlings indoors so early, in the depths of bitter winter, that the plants are flowering, or even fruiting, by the time it is safe to set them out in the garden. The effort is often wasted, however, since younger seedlings usually catch up with older ones once transplanted into the garden.)

The number of days it takes for a tomato to mature can only be a rough estimate, at best. Maturation of the heat-loving plants will be slowed down by a string of overcast days or prolonged cold spells, so that a variety said by a seed catalogue to ripen in 55 days may actually take 75 or 80. This is why gardeners in Canada, the Pacific Northwest, and other regions with short summers and cool temperatures grow early tomatoes more often than midseason ones; in their climates, early varieties actually ripen in the middle of the season at the soonest, and the real midseason varieties may never have time to ripen at all.

In general, the fruits of early varieties are the poor relatives of the tomato family. Nearly all of them are small and rather hard, and they usually lack the special succulence and flavor that come to a tomato only by staying attached to the vine for a long time.

If early varieties are not listed separately from midseason varieties in a seed catalogue, the gardener can pick them by the figure given for "days to maturity." Early types are those that are listed as ripening in 64 days or less. (The gardener should take into consideration whether the catalogue is regional; the "days to maturity" cited in a catalogue from New England may be many more than the same variety requires when grown

29

farther south.) Oddly, the word "early" in the name of the variety does not always signify earliness. Early Pak 7, for example, usually takes up to 80 days to ripen, longer than most midseason types; an obscure variety called Early Detroit takes about 78 days.

Early varieties should by no means be confined to gardens in regions with short growing seasons. In more temperate zones, where the last frost usually comes in April and the first frost in October or November, gardeners can ensure a steady supply of tomatoes by planting all three types—early, midseason, and late. This staggered harvest will also spare the gardener the risk of growing tired of eating tomatoes with the same flavor and texture all summer long.

Midseason tomatoes, often called main crop tomatoes in seed catalogues, usually take at least 65 days to mature under normal conditions. They are larger than early types, and usually juicier and softer. Late varieties take 80 days or more, and include all of the huge beefsteak types.

INDETERMINATE AND DETERMINATE VARIETIES

Gardeners may be perplexed when, reading through seed catalogues, they see tomatoes described as indeterminate or determinate, and may dismiss the labels as minor points they need not worry about. The difference is not complex, however. The branches of *indeterminate* plants keep growing and producing fruits until frost kills the plant. The branches of *determinate* plants stop growing and producing fruits when they reach a certain length. This means, obviously, that indeterminates are big plants, and determinates are small. (Some determinates are larger than most and are sometimes labeled with the vague, contradictory term "semi-determinate"; they are more accurately called large determinates.)

Whether a tomato plant has one growth habit or another is not a minor point, but an important one. Indeterminates and determinates produce entirely different kinds of crops: indeterminates set fruit all season long, so that the gardener has an extended harvest; determinates have a relatively concentrated fruit set, meaning the gardener may harvest tomatoes only for two or three weeks. This makes indeterminates the better choice for people who want a few fresh tomatoes all season long, while determinates are desirable for those, such as home canners, who want a large crop in a short period.

Production is not the only difference between the two types. Care of the plants and even the quality of the fruits are different.

Indeterminates are too large to support themselves, so they are usually staked, caged, or trellised. On stakes or trellises, they are sometimes pruned to two or three main stems. Such pruning results in larger, though fewer, fruits. Although indeterminates can be grown unsupported, of course, they will take up too much room in all but the largest gardens. And many of the fruits will rest on the ground, where they risk rotting and invasion by ants and other pests.

Indeterminates generally have more foliage than determinates. Insulated by a heavy cover of leaves, their fruits are less likely to be affected by extremes in temperature, and will not risk being scalded by the sun. And because they are insulated, they will ripen more slowly. This can improve taste. (The high leaf-to-fruit ratio is also beneficial for taste in another way: the more photosynthesis that takes place in the plant, the more sugars are produced for the fruits.)

Determinates, sometimes called bush tomatoes, are usually small enough to support themselves sufficiently and need no staking. For this reason, they are popular with commercial growers; they require less labor and take up little room when allowed to sprawl. And because their fruits ripen within a few days of one another, the plant's entire production can be harvested in only two or three pickings.

Determinates have a place in the home garden, too. Gardeners with little space may like them because they can be planted closer together, when staked, than indeterminates; home canners need them because all the fruits are ready for processing at the same time; and those who grow tomatoes in containers will find determinates the only practical choice.

A common misconception among gardeners is that all early varieties of tomatoes are determinate, while all midseason and late ones are indeterminate. This idea has arisen, perhaps, because determinates exhaust their fruiting before the end of the growing season. And their fruits, once set, usually do ripen faster than those of indeterminates, since they have less foliage

31

cover and are therefore not as insulated from the heat that speeds ripening. But determinates do not necessarily set their crops early in the season—many wait until late summer; and some indeterminates can start ripening fruit as early as 55 days from transplanting.

Within the determinate and indeterminate classifications are a few subdivisions:

Dwarf varieties are small, truly self-supporting plants that are usually grown in containers. Patio and the old Dwarf Champion are the best-known examples. They are determinates and have a strong, woody central stem.

Dwarf tomatoes were formerly known as "tree tomatoes." Confusion has arisen in recent years, however. Some gardeners use "tree tomato" as another name for climbing varieties. (It should be understood that no tomato variety actually climbs by itself, as peas or beans do, but must be tied to stakes or trellises; climbing tomatoes are simply indeterminates with remarkably vigorous growth.) Adding to the confusion is the fact that one climbing tomato is named Giant Tree. And compounding the problem even more are the merchandisers who advertise "tree tomatoes" that are not tomatoes at all, but an entirely different fruit called the tamarillo.

Miniature varieties, also determinates, are grown mainly for ornamentation. They are intended for container growing, and their vines are sometimes no more than 8 inches long. Minibel is one of the better-known varieties.

HYBRIDS AND STANDARDS

Before the 1940s, horticulturists developed new tomato varieties by selecting plants with certain desirable characteristics, saving the seed, replanting it, and continuing to select the most favorable plants. After a few years of such selection, undesirable qualities were eventually bred out. One variety could also be crossed with another, and their progeny put through the same selection process.

The last forty years, however, have seen the introduction of first generation hybrids. By definition, hybrids are the result of crossing two different parent varieties; but in the case of first generation hybrids (called F_1 hybrids for the sake of brevity), the parents are the greatly divergent offspring of a long line of breeding-stock tomatoes that have certain qualities. One, for example, may produce extra-large fruits; another may have a gene that makes it resistant to nematodes; another may resist verticillium or fusarium wilt, and still another may have skin elastic enough to resist cracking. All of these characteristics are carried in the two parents used to produce the final hybrid, and the seeds of that hybrid—the ones the gardener will buy—will produce tomatoes with all of the favorable characteristics of the ancestors. If seeds of the hybrid are saved by the gardener from the first season's crop, however, and planted the next season, segregation will occur: the plants will revert to their various original ancestors and yield as many as five or six different varieties. (Lest the gardener thinks this sounds like a good way to get six varieties for the price of one, he should know that these tomatoes will be suitable only as breeding stock, not as fruits anyone would wish to eat.)

Because F_1 hybrids come true to seed only once, they must be reproduced through hand pollination every year by seed companies. In normal pollination, the fruit-setting ovary of the tomato flower is fertilized by pollen from the flower's own anther. In hybrid pollination, however, the pollen-producing anther is removed from the tomato flower by hand, and the pollen of another parent plant is deposited, again by hand, on the stigma of the flower; this results in fertilization, and eventually in a fruit with all the characteristics of both parents, at least for the first generation. It is this time-consuming hand pollination that makes hybrid seed so expensive.

An understanding of the botanical technicalities of standards and hybrids will not help the gardener grow better tomatoes, but a knowledge of their differences will help him make an intelligent choice from the hundreds of varieties available.

Nonhybrids are called open-pollinated varieties, or, more commonly, standards. (Some catalogues simply identify them by the initials O.P. Other catalogues do not identify them at all; unidentified varieties are usually standards, since law decrees that hybrids must be labeled as such.) Standards include most varieties introduced before the 1970s, but also include some of the commonly grown later varieties. And development of new standard varieties still continues by plant breeders, although they are often overshadowed in the glare of promotion surrounding the hybrids.

Hybrids do indeed perform well in the garden. They have "hybrid vigor," a term that is often used in seed catalogues and should not be viewed by the gardener as mere hype. Hybrid vigor, known scientifically as heterosis, is a phenomenon in which the hybrid plant shows more favorable growth factors than either of its parents. And a tomato with several ancestors will have not just one or two favorable growth factors, but many, including disease resistance, good fruit size and shape, and high yields.

Yet, as with most things that seem too good to be true, there is controversy. Many people claim that breeding so many qualities into even a home garden tomato will affect its taste for the worse. Moreover, hybrid seed costs more to buy than standard seed, and must be bought again every year. And while hybrids may have resistance to the most common tomato diseases, they may be *less* resistant to other, less common ones, unlike the standards that have developed multiresistance through the long, slow process of selection. Hybrids often have to rely on dusts and sprays for protection. And they also depend heavily on large helpings of fertilizer.

Some critics see an ominous side to the rush to hybrids at the expense of older standards, and point darkly to the fact that most small seed companies are being swallowed up by ever larger conglomerates. Our seed supply, they say, has fallen into the hands of a few profit-hungry corporations. Articles are written to suggest that the main concern of a conglomerate's vegetable seed division is to breed hybrids that require large applications of the chemical fertilizers and pesticides manufactured by other divisions of the same conglomerate. The objective, they charge, is not the creation of genuinely better varieties, but of money-making ones.

Other people dismiss this as nonsense. They find the idea of a conspiratorial agribusiness that cares only for its own welfare, without a thought to the common good, simply naive.

Understandably, the innocent backyard tomato grower may want to keep politics out of his garden; gardening, after all, is one of the traditional ways people escape worries and cares. But if the gardener is concerned or confused, and suspects that he is being encouraged to grow hybrids mainly because it means more profits for a few giant conglomerates, he might choose to grow varieties of both types—hybrids for their undeniable attributes, and standards to help keep the demand for older varieties alive.

HEIRLOOM VARIETIES

Heirloom varieties are the true old-timers, most of which date from the nineteenth century, when tomatoes were becoming commonplace in North American gardens. Today, there is an ineffable mystique about them for many gardeners. Their evocative names—Brandywine, Mortgage Lifter, Garden Peach, Mammoth German Gold, and others—add to their fascination and make some gardeners all the more determined to find them and grow them.

Because they are old does not necessarily mean they are better, of course; after all, most have disappeared because they have been improved upon by the products of twentieth century plant breeders. A few, including Ponderosa, Yellow Pear and Yellow Plum, still are widely grown and are readily available through seed catalogues and at nurseries. But most are hard to find, if they still even exist. One problem is that a great deal of confusion surrounds many of the varieties. It seems there are more names than there are tomatoes; that is, one old variety may have been known by several different synonyms in various parts of the country and to different generations. (Ponderosa has been known as Grand Pacific, Majestic, Colossal, and Peak of Perfection, as well as several other appellations.) The gardener who goes in search of an heirloom he thinks is in danger of disappearing may not realize that it is not on the verge of extinction, but is merely being sold under another name.

Gardeners interested in old varieties may want to become members of the Seed Savers Exchange (Rural Route 2, Princeton, Missouri 64673), a preservation network that tries to keep alive hundreds of heirloom varieties of vegetables, including tomatoes. The organization has an information letter that describes its projects and publications; all the gardener need do to obtain one is send a self-addressed, stamped envelope.

DESCRIPTIONS OF VARIETIES IN THIS BOOK

On the following pages are descriptions of 325 varieties of tomatoes of all types. Nearly all are available to home gardeners, either through seed catalogues or at nurseries. The few that are not available were once popular, but have fallen out of circulation for one reason or another; they are included for the sake of reference.

One hundred of the most commonly grown varieties are described in detail, and are rated as excellent, good, or fair. (It is presumed that no truly poor varieties survive in the marketplace.) Like many of the cultivation techniques in part 3 of this book, the ratings are those of home gardeners, 2,000 of whom were asked, by questionnaire, to give their opinions of the varieties they have grown over the years. They judged each variety by its taste, how much fruit it yielded, and how well it resisted disease. While these judgments—one of them subjective and the other two objective—are recounted separately in the text, they are lumped together in the overall designation of a variety as excellent, good, or fair. As expected, opinions varied widely on taste, and descriptions of yields and disease resistance changed from region to region. But a general consensus emerged. One respondent's favorite variety was sometimes another's least favorite, but the fact that certain varieties were consistently rated by gardeners as winners—while others were consistently rated as also-rans—made it safe to assume that a general collective opinion prevails. That opinion, translated here as a one- to three-star rating, will give the reader an idea of which tomatoes are viewed with the most approval, by the most people, in North America today. And that approval, from everyday backyard gardeners, may be the most accurate and trustworthy appraisal of tomatoes that exists.

The ratings are indicated by stars placed before the name of each variety:

*** Excellent

** Good

* Fair

A notation at the end of each description lists the seed catalogues through which the variety is sold. Code numbers are keyed to the forty catalogues listed on pages 192–193. The catalogues—thirty-five from the United States and five from Canada—are not the only ones available to home gardeners, but they include those from the largest nationwide seed companies, several regional companies, and a selection of smaller companies that offer unusual varieties.

Besides the 100 rated varieties, 225 others are described briefly on pages 85–111. They, too, are keyed to the catalogues.

Each description contains the information the gardener needs in order to choose a variety:

Disease resistance is indicated in the way it usually is in seed catalogues and on nursery labels—with the initials of the diseases to which the tomato has inbred resistance listed in parentheses after the variety's name:

Better Boy (VFN)

Only the initial letters of the most common tomato problems are included. They are "V," for resistance to verticillium; "F" for fusarium (see page 179 for an explanation of "F_1F_2"); "N" for nematodes; and T for tobacco mosaic virus.

Resistance to less common diseases, such as early blight and gray leaf spot, is noted within the description of each variety.

Days to maturity (indicated at the end of the text in the detailed descriptions, and after the name of the variety in the brief ones) are given only as a measure of comparison among varieties, not as the time fruits will actually take to ripen. Real ripening time depends on several factors, including the number of cloudy days, the mean daily temperature and the tomato's location in the garden.

Fruit size, usually given as an average weight and diameter, also varies widely according to growing conditions. Like the number of days to maturity, the fruit size is given so that varieties may be compared.

Shape is described only when it is unusual, on the assumption that the finer differences between the shapes of tomatoes—globe vs. oblate, for example—are not important to home gardeners.

Color, like shape, is rarely mentioned unless the tomato is pink, yellow, or white, on the assumption that the gardener does not care what shade of red his tomatoes are.

Taste is described objectively as acid, sweet, or with an equal balance of both qualities. Subjective judgments are those of the home gardeners who were surveyed.

100 TOMATO VARIETIES, RATED AND DESCRIBED

*** Abraham Lincoln

Midseason to late

Standard; hybrid available

Indeterminate

Abraham Lincoln is an old beefsteak variety introduced by an Illinois seed farm called Buckbee's in 1923, around the time when tomatoes were becoming the country's most commonplace backyard crop. As beefsteaks go, the variety is unusual: its fruits are round and smooth, rather than ribbed and irregular; and its foliage has a certain bronze tinge, a characteristic that may lead first-time growers to believe that the plant is affected by some strange disease. Despite the odd color of its leaves, it produces bright red fruits—as many as nine to a cluster—that weigh between 1 and 2 pounds each. They are meaty, with few seeds, and usually free of cracks and seams.

The variety is thought of as a midwestern specialty, although growers in other parts of the country (with the exception of those with extremely short growing seasons) have reported successful harvests. Growers give the tomatoes high praise, describing their taste as mild, yet not bland.

Buckbee's seed farm was bought in later years by the R. H. Shumway company, now the main mail order vendor of this variety. Shumway's has developed a hybrid version of the Abraham Lincoln; it is said to be more prolific and more disease resistant. And although some old-timers insist that improved hybrid versions of old-fashioned tomatoes lack the taste of the originals, there is no evidence that the hybrid Abraham Lincoln is inferior to the original. Taste tests show that the two are indistinguishable. Days to maturity: 80. Catalogues 11, 31.

* Ace

Late

Standard

Indeterminate

Ace was developed in 1953 by the Campbell Soup Company as a processing tomato for soups and other tomato-based products. But it was also put to other uses: it was widely grown by commercial growers (primarily in California, but also in the East), and became one of the more common supermarket tomatoes of the 1950s. Paradoxically, it became extremely popular among home gardeners as well—proof, perhaps, that commercial varieties can overcome their terrible reputation if left to ripen properly on the vine instead of being picked green and shipped across state lines.

Home gardeners like Ace because its taste is pleasantly mild, with a good ratio of sugar to acid. Its performance, however, can be poor. While it often does well in parts of the West, producing smooth, thick-walled fruits weighing up to 8 ounces, it is described by gardeners in other parts of the country as inconsistent: yields are small and fruit does not always ripen. Days to maturity: 85. Catalogue 29.

Three standards and one hybrid, all with disease resistance that the original does not have, have been developed:

*Ace 55 (VF), a standard introduced in the mid-1960s, has fruits that are not quite as smooth as those of the original; and they have been rated "low in flavor" in taste tests. The variety's main recommendation is its resistance to verticillium and fusarium wilts. Determinate. Days to maturity: 80. Catalogues 4, 11, 22.

**Ace-Hy (VFN), popularly known as Hybrid Ace, was developed in 1977 in California. It is not only resistant to verticillium and fusarium wilts, but also to nematodes. It is slightly earlier than the other Aces. Determinate. Days to maturity: 76. Not sold through listed catalogues; available at nurseries in the West.

*Cal-Ace (VF), a standard, was developed from Ace 55 to set fruit in hotter climates. It yields more and smoother fruits, but, like Ace 55, has been rated low in taste tests. Large determinate. Days to maturity: 80. Catalogues 5, 12, 20.

*Royal Ace (VF), a standard, has larger fruits than the original, yet makes a more compact plant. Because it will not set fruit in high temperatures, it rarely is grown west of the Mississippi. Determinate. Days to maturity: 80. Not sold through listed catalogues; available at nurseries and at retail seed outlets.

** Atkinson (FN)

Midseason to late

Standard

Indeterminate

Developed at Auburn University in Alabama in 1967, this variety was bred for the special needs of gardeners in the Southeast: it is resistant to nematodes, which are a chronic problem for gardeners in that part of the country; the plant thrives in a humid climate; and its heavy foliage shades the fruits from the scalding southern sun.

Atkinson produces a heavy yield of 6-ounce fruits. Sugar and acid content is fairly high, so that gardeners describe the tomatoes as full-flavored. Even though the vine is indeterminate, enough fruits mature at once to make Atkinson a good choice for growers who want to can tomatoes for winter.

Atkinson is resistant to fusarium wilt as well as to nematodes. It also has tolerance for two common southern tomato diseases, early blight and gray leaf spot. Days to maturity: 75. Catalogue 15.

** Basket King

Early

Hybrid

Dwarf ornamental, Cherry

House plant enthusiasts, unwilling or unable to grow tomatoes in the backyard, are the main beneficiaries of this variety developed by Burpee's. Bred especially for containers, it is grown as much for its appearance as for its fruit: leafy branches, bearing clusters of cherry tomatoes, cascade over the sides of hanging baskets, window boxes, or flower pots.

The plants grow well in containers as small as 8 to 9 inches across. They are usually placed on patios or balconies, but will also thrive indoors in a window that receives a few hours of direct sunlight each day. Compact and neat, Basket King is one of the few container plants that provides both ornamentation and food. The fruits measure 1¾ inches across, and, like many cherry tomatoes, are thin-skinned and sweet. Days to maturity: 55. Catalogue 4.

** Beefsteak

Standard

Late

Indeterminate

The word "beefsteak" is customarily used to describe any tomato that weighs 1 pound or more and has meaty flesh with few seeds and little gel. But gardeners will also find a specific variety called Beefsteak listed in their catalogues—an old standard that formerly went by the name of Crimson Cushion. It is the tomato that many people speak of as "old-fashioned," the one they remember their grandparents growing, and they claim its full, sweet flavor sets the standard by which tomatoes should be judged. Innumerable other tomato growers, however, are not nearly so impressed. In a survey of home gardeners, Beefsteak scored highest in two categories: best and worst. Taste, after all, is highly subjective.

Its detractors describe Beefsteak as bland, because of its low acid content; they also find its flesh coarse and mealy. Some gardeners object to its appearance, although this may be because they have been conditioned by advertisers and commercial food growers to expect produce that is blemish-free and of a regular shape. Beefsteaks do not live up to this ideal: they are ribbed and irregular, sometimes appearing grotesquely misshapen, and often have an unsightly blossom end scar. They also have large cores, so that much of the flesh at the stem end is inedible.

In areas where the growing season is long, Beefsteak usually produces a good yield. Gardeners who live in areas with short growing seasons but who nevertheless are determined to harvest enormous tomatoes should plant another variety: Beefsteak is almost always a disappointment, since it rarely has time to ripen before first frost in the northern states and Canada.

Growers east of the Mississippi have more luck with Beefsteak than those in the West, where nights are too cool and the humidity too low. Beefsteak sets fruit best where night temperatures average about 60° or higher; but they will refuse to set fruit when day temperatures exceed 95°. And they grow best in areas with high humidity.

Western gardeners have reported success by planting Beefsteak next to a south- or west-facing wall which absorbs the sun's heat during the day and radiates it back at night; this warm microclimate helps prevent blossom drop.

Beefsteak is so vigorous that it is likely to grow uncontrollably under ideal conditions; for this reason, it is better supported by a cage than by a stake. Days to maturity: 82. Catalogues 2, 3, 7, 16, 22, 24, 26, 29, 31, 37.

Two improved varieties are earlier and, in most respects, more reliable. All have the solid flesh and sweet taste characteristic of beefsteak types:

Beefmaster (VFN), formerly called Beefeater, is an indeterminate hybrid with inbred disease resistance. Its fruits are larger, weighing up to 2 pounds. Days to maturity: 80. Catalogues 2, 3, 4, 6, 10, 12, 14, 15, 17, 20, 22, 25, 26, 29, 30, 31, 32, 37, 38, 39.

***Super Beefsteak** (VFN), an indeterminate standard developed by Burpee's, is also disease resistant. The fruits are as large as those of the original Beefsteak, averaging about 1 pound, and are smoother, with fewer ridges. Days to maturity: 80. Catalogues 4, 34.

Two other varieties are not particularly large, but are called beefsteaks because of their meaty character. Both are much earlier than the larger types:

Bush Beefsteak, a standard, is a compact determinate that has an especially heavy yield, although its fruits average only about 8 ounces each. Its short maturation time makes it a common choice for gardeners in Canada and the northern states. Days to maturity: 62. Catalogues 32, 36, 37, 39.

Prime Beefsteak (VF), a determinate hybrid developed by Ferry-Morse, has 8 to 10 ounce fruits that are more regular in shape than the larger beefsteaks. Days to maturity: 70. Not sold through listed catalogues; available at nurseries and at retail seed outlets.

*** Better Boy (VFN)

Midseason

Hybrid

Indeterminate

There is no question that Better Boy is the favorite home garden tomato in North America today. Introduced in the early 1970s by the Petoseed Company of California, it was bred to set fruit at a wide range of temperatures, so that it performs as well for gardeners in hot climates as it does for those who live in the northern states and Canada. (A number of varieties are touted in catalogues as "adaptable to all areas," but few live up to the promise.)

There are other reasons for Better Boy's status as the most successful hybrid of the last decade: it consistently produces a huge yield of large, juicy, shapely fruits that have the taste people expect of a real tomato; and it is relatively trouble-free to grow, having been bred to resist the problems that are most likely to affect tomatoes: verticillium wilt, fusarium wilt, and nematodes.

Fruits usually weigh from 12 ounces to 1 pound, but often grow larger. They have a good balance of sweetness to acidity, and contain more juice than many varieties, yet are firm enough to be described in catalogues as "good for slicing." They have rather large cores, though not so large as those of the beefsteak varieties.

Home gardeners who were surveyed were virtually unanimous in their approval of Better Boy, and many claimed it was the most vigorous variety they had ever grown. Disappointments were few: some growers found Better Boy susceptible to blossom end rot, but overcame the problem by giving the plants extra calcium and regular, even watering; a grower in Connecticut described the tomatoes as not so much juicy as watery. But these irregularities occur so rarely that gardeners can be almost certain that Better Boy will live up to the sometimes extravagant claims made for hybrid tomatoes in the seed catalogues.

Given the right amount of water, sun, and fertilizer, the plant can grow as tall as a house; obviously, strong stakes or cages are needed. Days to maturity: 72. Catalogues 2, 3, 4, 5, 6, 7, 10, 12, 14, 15, 19, 20, 21, 22, 24, 25, 26, 28, 29, 31, 32, 33, 34, 35, 36, 37, 38.

** Big Boy

Midseason to late

Hybrid

Indeterminate

Big Boy burst upon the scene in 1949—a new-fangled hybrid with a catchy name, the creation of the nation's largest seed company, W. Atlee Burpee. It arrived just as the "victory gardeners" of World War II were becoming curious about new hybrid vegetables, and it promised king-sized tomatoes with yields 20 percent higher than those of the best open-pollinated varieties. And Burpee's, in a brilliant stroke of marketing, gave it a name that was at once memorable and descriptive, at a time when the names on tomato seed packets had no more zip than Stone, Marglobe, or Rutgers.

Burpee's had introduced the first hybrid tomato available to the general public, Fordhook, in 1945. It was an early variety with a higher yield than Earliana and Valiant (then the two most commonly grown commercial varieties in the East), and matured more quickly, as well. While Fordhook was successful for its time, it did little to excite the imagination of home gardeners, who were more interested in large-fruited midseason varieties. Eventually, Fordhook was superseded by two other Burpee early varieties, Burpeeana and Big Early, and was dropped altogether in 1960.

Yet, in the 1980s, it is Big Boy that is being superseded. The trend it started in nomenclature—Better Boy, Early Girl, Wonder Boy, and Ultra Girl are only a few of its namesakes—is old hat, and horticultural scientists have bred specific disease resistance into midseason hybrids so that they are much less troublesome to grow.

Nevertheless, Big Boy remains the favorite of countless home gardeners. Some grow it out of habit, but many other gardeners prefer it because they say its taste is better than that of the new hybrids.

The fruits of Big Boy average about 12 ounces, but often weigh 1 pound or more. They are meaty, bright red, and are exceptionally flavorful, bursting with what people call "real tomato taste." However, as often as Big Boy is praised for its taste, it is criticized for its low yield—an irony, considering it made its initial splash as a high-yielder. It also testifies to the extremely heavy production of the newer hybrids.

Yields can be improved by growing the large indeterminate vine in a cage. Leaving the plant unpruned will result in more, though slightly smaller, fruits. Gardeners in areas with short growing seasons—particularly the Pacific Northwest and Canada—should try another variety, since Big Boy rarely ripens in those areas before first frost.

Big Boy may have been improved upon by later hybrids—and no challenge was ever more direct than that of the obviously named Better Boy—but it still is the sentimental choice of a huge number of American home gardeners. Days to maturity: 78. Catalogues 1, 2, 3, 4, 7, 9, 10, 12, 15, 19, 21, 22, 24, 25, 26, 28, 29, 30, 32, 33, 34, 35, 37.

*** **Big Pick** (VF$_1$F$_2$NT)

Midseason

Hybrid

Indeterminate

Big Pick is one of the new breed of supertomatoes of which Better Boy is the best-known example: a high-yielding, large-fruited hybrid with built-in disease resistance, and bred especially for the home garden. Big Pick was introduced in the late 1970s, almost a decade after Better Boy, and is gaining favor because of its superior disease resistance. Besides resisting verticillium wilt, races 1 and 2 of fusarium wilt, nematodes, and tobacco mosaic virus, it has good tolerance for early blight.

Fruits usually weigh about 8 ounces, but, like Better Boy, often grow to 1 pound or more when conditions are ideal. The plant sets fruit in most temperatures. Fruits are round, smooth, fairly firm, and do not have the green shoulders common to many large varieties. The flavor is praised by gardeners as exceptionally good—neither too sweet nor too acid, but with a good balance of both qualities.

Since the vine is likely to grow very tall, caging or staking is necessary. Days to maturity: 70. Catalogues 10, 15, 26, 29, 32, 37.

** **Big Set** (VF$_1$F$_2$N)

Midseason

Hybrid

Large determinate

Like a number of hybrids intended primarily for commercial growing, Big Set was developed as an improvement on the Walter variety. (Walter is a venerable Florida high-yielder that is resistant to races 1 and 2 of fusarium wilt.) Big Set, introduced from California in 1970, was bred not only to have more disease resistance—it is resistant to verticillium wilt and nematodes as well as to fusarium—but also to produce fruits more uniform in size than those of its Walter parent. Like Walter, it is mainly grown commercially, but has enough flavor when picked dead ripe to make it a good home garden variety as well. Many home gardeners describe Big Set as especially succulent.

Seed catalogues from areas as diverse as Texas, eastern Canada, the Pacific Northwest, and the mid-Atlantic states claim that Big Set is a first-class performer in their climate zones. Indeed, it does set fruit at a wide range of temperatures, and is earlier than most midseason varieties. Fruits have the firmness characteristic of commercial varieties, and are smooth and large, weighing 8 to 9 ounces and measuring about 3 inches across; they are fairly resistant to blossom end rot, cracking, and catfacing. Yields are good, with most of the fruits becoming ripe at the same time under the protection of a heavy foliage cover.

The vine is determinate, and can be grown unstaked and unpruned without sacrificing fruit size. It is larger than most determinates, however, so that many gardeners prefer to support it on short stakes or in cages. Days to maturity: 65. Catalogues 9, 24, 28, 34, 40.

* **Bonny Best**

Early to midseason

Standard

Indeterminate

Bonny Best has survived as a home garden variety since early in the century because of its robust acidity and its quickness to ripen. It grows well in most parts of the country, and matures as early as 60 days from transplanting in the Southwest and West to 75 days in the northern states and Canada. Fruits have a pleasantly sharp flavor, but are rather small, averaging about 4 ounces each. In hot climates, the fruits may be even smaller.

Gardeners rate Bonny Best higher for its taste than for its yield, which is often disappointing. Disease resistance is not particularly good, either, especially in humid parts of the Southeast. Northern gardeners generally report better performance from this variety; southern gardeners say they have increased yields and prevented sunscald by growing the plants in a cage.

Bonny Best is still known by many old-time gardeners as John Baer. Whoever Mr. Baer was, he was not the originator of the variety. Bonny Best was the find of one George W. Middleton, of Jeffersonville, Pennsylvania, who saved seed from a superior plant he found growing in a field of Chalk's Early Jewel, a variety common around the turn of the century. A seed company introduced the seed from Middleton's plant in 1908, and Bonny Best—which has been known over the years in various regions as John Baer, Early Marketeer, Red Bird, and Redhead—has been in demand ever since. Days to maturity: 70. Catalogues 7, 14, 18, 24, 25, 32.

** **Bonus** (VFN)

Midseason

Hybrid

Large determinate

Southern gardeners describe Bonus as "good and sweet." Its heavy foliage cover and ability to set fruit at high temperatures make it popular from the Carolinas to Texas. It also is available in the mid-Atlantic states, though its 75-day maturation time makes it a risky choice for growers north of Pennsylvania and New Jersey.

Introduced in 1964, Bonus was one of the first multidisease resistant tomatoes. Unlike many other resistant varieties, however, it is a determinate vine—which means it can be grown unstaked. (Like its cousin, Big Set, Bonus is larger than most determinates, so that many gardeners prefer to grow it in cages or on short stakes.)

Yields are high, with fruits weighing 6 to 7 ounces and

measuring 2¾ inches across. Although some catalogues claim the tomatoes are resistant to catfacing and cracking, many growers complain that the fruits often have deep radial cracks. They are generally complimentary about the flavor, however, which is slightly sweeter than that of most tomatoes. Days to maturity: 75. Catalogues 26, 28.

** Bradley (F)

Midseason

Standard

Large determinate

This pink tomato was developed at the University of Arkansas in 1961 as a commercial variety that would resist the fusarium wilt that its two commonly grown predecessors, Gulf State Market and Pinkshipper, were subject to. (It takes its name from Bradley County, the center of commercial tomato growing in Arkansas.) But it soon was in demand by home gardeners, especially in Arkansas and Louisiana, because it lacked the rock-hardness of other commercial varieties, and because it had the characteristics that make pink tomatoes desirable—meatiness and mild, sweet flavor.

Bradley is rather like a smaller version of the classic pink tomato, Ponderosa, with the same green, prominent shoulders of that variety. And, like Ponderosa, it is subject to the problem that plagues many older standards—radial cracking. Fruits are globe-shaped, weighing 7 to 8 ounces and measuring about 3½ inches across. Bradley is best grown in cages or on stakes and pruned to two stems; even when pruned, the heavy foliage will offer good coverage for the fruits. Days to maturity: 74. Catalogue 10.

Two more recent Arkansas-developed pinks, Traveler and Traveler 76, have replaced Bradley in commercial production, and in some home gardens, because they are less subject to cracking. Many gardeners have stayed with Bradley, however, because its taste is undeniably better and because they find the Travelers too firm.

*Traveler (F), sometimes called Arkansas Traveler, is an indeterminate with resistance to cracking and fusarium wilt. It should be grown on a stake and pruned to two stems in order to ensure good fruit size. Days to maturity: 76. Not sold through listed catalogues; available at nurseries in Arkansas and Louisiana.

*Traveler 76** (F) is earlier than both Bradley and Traveler, and has still more crack resistance. Like Traveler, it should be pruned to two stems. Days to maturity: 72. Catalogue 38.

* Bragger

Late

Hybrid

Indeterminate

Bragger is usually described as the variety for gardeners "who want to grow the biggest tomato on the block." But those to whom the gardener is bragging should be aware that the tomato is big not because of superior cultivation on anyone's part, but merely because Bragger was bred to be enormous. Fruits weigh as much as 2 pounds or more, and a single slice is likely to overhang the sides of a sandwich. A beefsteak type, the tomato is irregularly shaped and has solid flesh with small seed cavities. Catalogues tout it as "big, meaty, and flavorful"; many home gardeners describe it as coarse and bland, if undeniably huge.

Yields are relatively low, as they are for most beefsteaks. (Growers report increased yields by pruning the vine to one stem and supporting it in a cage.) On the positive side, Bragger is generally free of the cracking and splitting that occur in many large varieties. Days to maturity: 85. Catalogues 2, 5, 9, 10, 11, 26, 29, 36, 37.

*** Brandywine

Midseason to late

Standard

Indeterminate

In the age of the supermarket, and its pale commercial produce, Americans may have idealized the tomato in memory. Real fanciers are ever on the search for the perfect tomato, and expect to find it among the older, more obscure varieties—German pinks or Polish reds, Mexican yellows or Kentucky purples. Some growers, most of them in the Midwest, stopped looking once they found Brandywine.

Even if Brandywine's status as the perfect tomato is open to question, it is certain that no other variety has developed such a mystique. No one seems to know where it came from or how long it has been around. It is not commercially available, and may never have been; it has been kept alive because its seeds have been passed from one gardener to another. And all who grow it agree that its richness and succulence have never been equaled.

But is Brandywine actually a distinct variety? Is it even an old one? Most gardeners have the idea that it dates from the nineteenth century, but no one can be found who has grown it for more than fifteen years. The old-timers who passed the seeds on have died. (A 97-year-old gardener in Ohio has grown it only since 1970, but says the seeds were given to him by a woman, now dead, who claimed they had been passed down in her family for more than a hundred years.)

Without doubt, Brandywine is very similar to some old beef-steak varieties—in particular, to two that are still commercially available, Climbing Trip-L Crop and Giant Tree (not to be confused with the so-called "tree tomato" advertised in Sunday supplements; the "tree tomato" is not a tomato at all, but an entirely different plant whose fruits happen to be red). Brandywine, like Climbing Trip-L Crop and Giant Tree, is a tall-growing, potato-leaved type (its leaves are not divided, as are most varieties', but look like those of the potato), and produces large, slightly rough, deep-pink or purplish fruits. Growers insist that Brandywine is meatier than either, and is indescribably more flavorful. But a tomato probably always tastes better if it is hard to come by.

Brandywine is most commonly grown in the Midwest, and is best known in Indiana. Perhaps a gardener there discovered a genetic mutation among a crop of Giant Tree or another potato-leaved variety, saved the seeds, and named the find Brandywine.

However the variety originated, it has survived, even though no one sells it. Most gardeners will have access to its seeds only if they become members of a seed exchange group such as Seed Savers Exchange (page 35), or if they are lucky enough to be acquainted with a gardener who grows this elusive "perfect tomato." Days to maturity: 85.

*** Campbell 1327 (VF)

Midseason

Standard

Large determinate

When Joseph Campbell started canning tomatoes back in 1869, the process was simple: he had to make do with the fruits nature gave him and seal them in a can. Today, plant breeders design tomatoes to have characteristics desirable for canning, catsup-making, soup, or sauce. Campbell's company, later called Campbell Soup, even established its own tomato breeding division in the 1940s to tailor tomatoes precisely to the company's needs.

One of the most successful of these processing tomatoes is Campbell 1327, first grown by the company in 1962 and made available to the seed trade soon thereafter. It superseded the most common processing variety of the time, Improved Garden State, and an earlier company development, Campbell 146, because it was earlier and had a higher yield; most important, it was highly crack-resistant, an essential trait for factory-bound fruits that were not picked until fully ripe. The tomato quickly became popular with home gardeners because it had good flavor, ripened uniformly, and resisted disease.

Fruits weigh about 7 ounces and are slightly flattened. They are juicy, and taste more acid than sweet, with a pleasant tang. Another advantage is that they are fairly soft, unlike most commercial types, and even show a blossom end scar that makes them look more like an old-fashioned tomato than a product of modern plant breeding.

The large determinate plants provide good foliage cover, and resist verticillium and fusarium wilts. They perform best in the northeastern part of the U.S. (as far west as the Dakotas and as far south as Kentucky), and in the southern parts of Ontario and Quebec. Days to maturity: 70. Catalogues 12, 22, 32, 34, 38.

*Campbell 19 (VF) was developed in 1965 as an even more crack-resistant variety. Fruits are rounder, firmer, and larger, weighing 8 to 10 ounces. But their taste is inferior to that of 1327. Campbell's no longer grows the variety, but it still is available to home gardeners. Large determinate. Days to maturity: 75. Catalogue 32.

**Campbell 146 (F) predates both varieties. It was bred in the

51

late 1950s as the first crack-resistant determinate, but was superseded by 1327 and 19 because of its lateness and relatively
low yields. Its taste, however, is superior: it still is used as the
standard for flavor at Campbell's. Fruits weigh about 7 ounces.
Days to maturity: 78. Not sold through listed catalogues; available
at nurseries.

* **Caro Red**

Midseason

Standard

Indeterminate

Tomatoes provide more vitamins and minerals in the diets of
North Americans than any other fruit or vegetable. Yet the tomato
itself is not nearly as nutritious as a number of other foods: in
the list of commonplace fruits and vegetables, the tomato ranks
only 16th as a source of vitamin A and 13th as a source of
vitamin C. Its number-one ranking in the nutrition tables is
explained by the public's heavy consumption of tomatoes, not
only as a fresh food, but also as an ingredient in processed
products. (Broccoli is the most nutritious commonplace food, but
ranks only 21st in its contribution of nutrients to the diet because
it is not eaten so often; tomatoes are consumed almost daily, in
everything from pizzas to Bloody Marys.)

A few varieties of tomatoes that are much more nutritious
than others have been developed. Because they are rather poor
specimens in every respect except nutrition, they are of interest
mainly to gardeners who want to supplement their vitamin A
or C intakes through natural foods rather than through vitamin pills.

Caro Red was bred in 1958 at Purdue University. Researchers
were not striving for a high-vitamin tomato, but simply for one
that resisted septoria leaf spot. They found that some of the
offspring of their crossbreeding experiments, with the wild
Lycopersicon hirsutum (hairy tomato) as a parent, had an unusual
orange-red color. Much of the lycopene, or red pigment, of these
progeny had been replaced by a yellow-orange pigment, betacarotene. (Tomatoes naturally have a small amount of betacarotene—about 4 or 5 percent of the pigment—but a much higher

52

percentage of lycopene.) The betacarotene content was about 50 percent, in fact; and since betacarotene is the source of provitamin A (the substance that is converted into vitamin A when ingested), Caro Red had ten times the amount of provitamin A as other tomatoes.

The extremely high nutritive value of Caro Red is, then, a by-product of its odd color. This is not necessarily true of other orange tomatoes, or of yellow ones; their color comes from other carotene pigments that have no provitamin A activity.

The vitamin content is the only reason a gardener might choose to grow this variety. A single Caro Red tomato supplies from 150 to 200 percent of the minimum daily requirement of vitamin A. Its taste, however, is poor; some people describe it as musky. Its texture is mealy. And it offers only a scanty crop of medium-sized fruits. Days to maturity: 78. Catalogue 3.

Two other high-vitamin tomatoes are sold:

*Caro Rich, an improved version of Caro Red, has a higher yield and a slightly better taste. Its fruits are the same size as those of the original—5 to 6 ounces—and are slightly flattened. They are less meaty, however. Vines are large determinates. Days to maturity: 80. Catalogue 32.

**Doublerich, developed at the University of New Hampshire in 1953, was purposely bred to have twice the vitamin C content of most tomatoes, in an effort to make them as nutritious as citrus fruits. It tastes no more acid than normal tomatoes, and is bright red in color. Fruits weigh 4 ounces and are quite firm. Because they ripen evenly, they are ideal for canning. (Tests by the USDA show the vitamin C content remains high even after the canned tomatoes have been stored for a year.) Determinate, standard. Days to maturity: 60. Catalogue 3.

** Celebrity (VF$_1$F$_2$NT)

Midseason

Hybrid

Large determinate

Introduced in 1984, Celebrity has lived up to its name by stirring up more publicity than any tomato since Better Boy. It announced itself as the first bush type with resistance to all of the major tomato problems—verticillium wilt, races 1 and 2 of fusarium wilt, nematodes, and tobacco mosaic virus. And because it sets fruits in all climate zones, from Canada to the southern United States, it won an All America Award, the first tomato to do so since Floramerica six years before. (All America Selections is an educational, nonprofit organization that evaluates new seed-grown vegetables and flowers. Previous award-winning tomatoes that are still available were Floramerica in 1978, Small Fry in 1970, Spring Giant in 1967, and Jubilee in 1943.)

Like Floramerica, Celebrity is a determinate, but has higher yields, producing a heavy crop of medium-sized tomatoes all season long. Although firm, the 7-ounce fruits have thin skin—a pleasant change from many of the home garden hybrids. While their flavor is fairly robust, it is not as exceptional as much of the new variety's publicity suggests.

The strong vines can be grown as a bush, but will take up less room in cages or on short stakes. Days to maturity: 70. Catalogues 3, 4, 6, 9, 10, 14, 15, 18, 20, 21, 23, 25, 26, 28, 32, 33, 34.

* **Chico III** (F)

Midseason

Standard

Determinate, Plum

The Chico varieties, the first of which was developed in Texas in 1961, are among the most widely grown processing tomatoes in the world. Bred for machine harvesting, they provide commercial growers with huge yields of small, pear-shaped fruits that end up on supermarket shelves as tomato paste, sauce, juice, or catsup.

Chico III is available for growing at home as an alternative to the more familiar paste variety, Roma (page 74). Some of the qualities that make it superior to Roma for commercial growing also appeal to home gardeners: it has higher yields, better disease resistance and more uniformly sized fruits (a consideration for home canners). And the fact that Chico III is hard and tough-skinned is of no consequence when the tomatoes are peeled and cooked down into a paste or sauce. The fruits lack the rich flavor of Roma, however, and are not nearly as good for eating fresh or using in salads.

Fruits weigh about 2 ounces and are 2½ inches long. They are borne on small, compact vines that are resistant to fusarium wilt and gray leaf spot. The plants are particularly well adapted to hot climates. Days to maturity: 75. Catalogues 19, 28, 30.

Royal Chico (vf) has larger fruits and more disease resistance. Its hard, thick-skinned fruits weigh about 3 ounces. Like Chico III, it has higher yields than Roma. Days to maturity: 78. Catalogues 12, 18, 23, 34, 39.

** Climbing Trip-L Crop

Late

Standard

Indeterminate

Its name notwithstanding, Climbing Trip-L Crop does not climb—at least not in the way of beans or peas. It takes its name from the vigor of its indeterminate vine, which can easily reach 15 feet or more by the end of a growing season. If trained to a trellis against the wall of a house, the vine will appear to have climbed to roof level or higher.

The "Trip-L Crop" in the variety's name is correct in its suggestion of huge yields. Gardeners can harvest as many as 2 or 3 bushels of tomatoes from a single plant. The dark pink fruits are large, averaging 1 pound or more. One slice measures 4 to 5 inches in diameter. Because the mild-tasting fruits are particularly meaty and have few seeds, they are often touted in catalogues as good for canning, although most canners might find them too large.

Climbing Trip-L Crop is one of the old-fashioned potato-leaved varieties. The shape of its leaves has no significance, but is a characteristic of some of the older beefsteak types. Days to maturity: 90. Catalogues 2, 3, 10, 12, 15, 25, 26, 28, 31.

** Delicious

Midseason to late

Standard

Indeterminate

Delicious is typical of the new, improved beefsteak varieties, producing rounder, smoother fruits that weigh 1 to 2 pounds. But this variety has a knack for yielding the occasional marvel— a tomato that keeps growing to enormous size. Clarence Dailey of Monona, Wisconsin harvested a Delicious in August 1978 that weighed in at 6 lbs. 8 oz. It earned a place in the Guinness Book of World Records as the largest tomato ever grown. The previous record, too, was held by a Delicious—4 lbs. 4 oz.

Introduced by Burpee's in 1964, Delicious is a cross between the yellow Jubilee and a large-fruited pink. Nevertheless, its color is fire-engine red. Gardeners praise its taste, saying it has more tang than most beefsteaks. The fruits also have smaller cores and smaller blossom end scars. Yields, however, are often poor; they can be increased by growing the plants in a cage and leaving them unpruned. Days to maturity: 78. Catalogues 2, 4, 6, 11, 12, 38.

* Earliana

Early

Standard

Indeterminate

The origin of Earliana, which dates from around 1900, is uncertain. One account says it was painstakingly developed by a farmer who was determined to market the first tomatoes of the season; another says it was born as a genetic mutant from a plant of the Stone variety. In any case, Earliana became a favorite early in the century.

Commercial growers rarely grow Earliana nowadays, because it has been superseded by a number of newer varieties with higher yields and better disease resistance. But home gardeners who prefer mild, subacid tomatoes still grow it for its mellow taste, bright red color, and quick maturity. Fruits weigh 4 to 5 ounces and are borne in clusters on open, spreading plants; they sometimes do not ripen evenly, and show green streaks and mottling at the stem end. Earliana is grown by gardeners in all parts of North America, but performs best in the West. Days to maturity: 62. Catalogues 9, 14, 25, 28, 29.

** Early Cascade (VF)

Early

Hybrid

Indeterminate

Early Cascade is one of the new hybrids that has replaced Earliana. Its taste, gardeners agree, is as good, if not better, than that of the older earlies. And it has exceptionally high yields, with clusters of tomatoes cascading from a vigorous indeterminate vine.

Fruits weigh about 5 ounces, ripen evenly, and are produced in abundance into the fall. The vines should be staked or caged. Days to maturity: 60. Catalogues 9, 10, 12, 15, 18, 20, 24, 25, 28, 29, 31, 34, 36, 37, 38, 40.

** Early Girl (v)

Early

Hybrid

Indeterminate

Home gardeners in the northern states, the Pacific Northwest, and Canada rate Early Girl as their favorite early tomato because of its reliability, saying it has consistently high yields and good disease resistance. And its taste is excellent, with a richness that is often lacking in early varieties. Gardeners in the South also rate Early Girl high for taste, but say the plants are susceptible to blight diseases in hot climates.

The meaty, crack-resistant fruits weigh 4 to 5 ounces and are produced throughout the growing season. Vigorous plants are leafy enough to offer good protection from sunscald, and are best staked or grown in cages. Days to maturity: 58. Catalogues 2, 3, 4, 6, 10, 12, 14, 15, 22, 23, 24, 26, 28, 33, 36, 37.

*** Early Pick

Early

Hybrid

Indeterminate

This tomato was developed by Burpee's as an improvement on their Big Early variety. Vigorous vines set fruit in a wider range of temperatures and also have more disease resistance. Burpee's especially recommends Early Pick for the West Coast, where low night temperatures often impair the fruit-setting ability of other varieties. Globe-shaped fruits average about 7 ounces, but can grow up to a pound each. They are solid and juicy, with a rather mild flavor, and ripen evenly all the way to the stem. Days to maturity: 62. Catalogue 4.

** Fantastic

Midseason

Hybrid

Indeterminate

Because Fantastic has extremely high yields and sets fruit in a wide range of temperatures, it is one of the most widely grown tomatoes in North America. Its meaty, 8-ounce fruits are borne profusely in clusters all season long, well protected by a heavy cover of leaves.

Most home gardeners rate Fantastic's yield as exceptional and its taste as good; but some complain that it succumbs too easily to disease. Days to maturity: 70. Catalogues 2, 5, 6, 9, 12, 14, 22, 26, 28, 29, 30, 32, 33, 34, 37, 38, 39.

***Super Fantastic** (vf) is a newer, improved version that will almost certainly replace Fantastic within a few years. It has the same high yields, with the advantage of resistance to verticillium and fusarium wilts. Its fruits are slightly larger, averaging about 10 ounces. Gardeners who have grown it rate it as one of the best tomatoes available. Days to maturity: 70. Catalogue 15.

* Fireball

Early

Standard

Determinate

Fireball was bred for gardeners with short growing seasons. It produces a heavy crop of tomatoes for a few days early in the season, and then is spent. Its firm, 4- to 5-ounce fruits are borne in clusters, and are typical of early types in that they are rather weakly flavored.

The plants produce best when left unstaked and unpruned, yet are small enough to plant as close together as 1½ feet—an advantage for gardeners with little space. Because the vines have sparse foliage, however, some gardeners prefer to grow them in cages; this crowds the leaves together and helps to prevent the sunscald that often affects this variety. Days to maturity: 60. Catalogues 7, 13, 14, 32.

* Flora-Dade (VF$_1$F$_2$)

Midseason

Standard

Determinate

This tomato was developed in the mid-1970s for the commercial growers of Dade County, Florida, after verticillium wilt was found to be reducing their yields by as much as 30 percent. It not only offered verticillium resistance, but could be mechanically harvested because of its concentrated fruit set and jointlessness (ability to be picked stem-free). Its superiority to the other two commonly grown Florida commercial tomatoes, Walter and Florida MH-1, led to its becoming one of the most commonplace supermarket tomatoes in the East by the 1980s.

Flora-Dade fares better in the home garden than it does in commercial fields, since its 7-ounce fruits develop better flavor when left to ripen fully on the vine. They are hard, however, and are produced for no more than 5 or 6 weeks during the growing season.

The plants have a heavy foliage cover to protect fruits from sunscald, and are resistant to gray leaf spot as well as to verticillium and fusarium wilts. Days to maturity: 78. Catalogues 5, 34.

* **Floramerica** (F₁F₂)

Midseason

Hybrid

Determinate

Tomato gardeners have found Floramerica one of the biggest disappointments of recent years, saying it has not lived up to its promise. It seemed to have everything going for it when it was introduced in 1978: developed by the University of Florida specifically for the home garden, it lacked the gratuitous qualities of commercial varieties; it won an All America Award, attesting to its adaptability to climate zones as diverse as Florida and Canada; and it had resistance to five diseases and tolerance for five more—an impressive feat on the part of the plant breeders. But for all its attributes, gardeners find its taste decidedly bland, and suspect that the breeders concentrated on disease resistance and adaptability at the expense of flavor.

Fruits weigh 8 to 12 ounces and measure 3 to 4 inches across. If the determinate vines are left unstaked and unpruned, the fruits will maintain their size throughout the growing season. Some gardeners report a shorter harvest of larger fruits, weighing up to 1 pound, by training the vines to one stem and fertilizing them heavily.

Although Floramerica is said to have yielded well in its trial plantings in the 1970s, many gardeners find its yields only fair.

Its ability to stand up to disease, however, is excellent. Floramerica is resistant not only to the two most common races of fusarium wilt, but to gray leaf spot and two other diseases that occur in humid areas—gray leaf mold and crown rot. It also has tolerance for verticillium wilt, early blight, tobacco mosaic virus, blossom end rot and bacterial fruit rot. Days to maturity: 72. Catalogues 1, 2, 3, 4, 5, 6, 9, 10, 11, 13, 14, 17, 19, 20, 21, 22, 24, 25, 26, 28, 29, 31, 32, 33, 34, 36, 37, 38.

*** Florida Petite

Early

Standard

Dwarf ornamental, Cherry

As container varieties go, Florida Petite is a breakthrough—the first dwarf ornamental that is small enough to grow in a 4-inch flower pot. Plant breeders at the University of Florida worked ten years to develop a variety that would grow well indoors on a window sill, and Florida Petite, introduced in 1983, is the happy result.

Cherry-sized tomatoes are borne on a tiny, compact plant that grows only 6 to 8 inches tall and spreads the same distance. About 25 fruits, 1½ inches across, are borne per plant. Their taste is typical of cherry tomatoes, but most gardeners prize Florida Petite as much for its attractiveness and novelty value as for its edibility.

Although seedlings are increasingly available at nurseries, many gardeners prefer to sow seed directly in a pot, using sterilized potting soil. The plant will bear fruit in only 60 days from seeding. (Transplanted seedlings mature in 40 days.)

The more direct sunlight the plant receives, the better it will flower and fruit. For this reason, it is best placed in a south-facing window. Alternatively, it can be grown outdoors in window boxes or on the patio. Days to maturity: 60 (from seed). Catalogues 10, 17, 32, 37.

** **Heinz 1350** (vf)

Midseason

Standard

Large determinate

The H. J. Heinz Company developed Heinz 1350 in the 1950s as a high-yielding tomato that would not crack or split when trucked to the company's processing factories. When it was released to the seed trade, it quickly gained a reputation as one of the best tomatoes for canning whole, although Heinz had bred it primarily for use in catsup or sauce. The tomato's superiority for whole-pack canning may have been largely assumed on the part of gardeners because of its name, but it does indeed have a number of qualities that make it a good canner: the 6- to 7-ounce fruits are firm, richly colored, uniform in size, coreless, and are produced in abundance within a fairly short time. Their only drawback is that they usually do not ripen evenly all the way to the stem.

Heinz 1350 is the only one of the Heinz varieties good for eating fresh, and is a favorite with gardeners who prefer tomatoes that taste more acid than sweet.

Large determinate vines set fruit in a wide range of temperatures, and should be left unstaked and unpruned to produce the highest yields. Days to maturity: 72. Catalogues 1, 4, 7, 9, 13, 14, 17, 25, 30, 31, 32.

Four more Heinz processing varieties are available to home gardeners:

*****Heinz 1370** (f) was bred to be suitable for machine harvesting. It is firmer and rounder than the slightly flattened 1350. Days to maturity: 77. Catalogues 32, 34.

*****Heinz 1409** (f), like 1370, was developed for machine harvesting, but is slightly larger and earlier. Days to maturity: 75. Catalogue 5.

** **Heinz 1439** (vf) is even better for canning than 1350, since it ripens all the way to the stem and has a more concentrated production of fruits. The 6-ounce tomatoes are protected by a heavy cover of leaves. Days to maturity: 70. Catalogues 10, 26, 29, 32.

*****Heinz 2653** (vf) is a paste variety, used only for cooking or canning whole. Plum-shaped fruits weigh about 2½ ounces and are very firm. While they cook down into a paste more quickly than Roma, San Marzano, and other home garden paste varieties, they are not as flavorful. Days to maturity: 68. Catalogue 18.

** Homestead 24 (F)

Midseason to late

Standard

Determinate

Homestead's evocative name leads some gardeners to assume it is an old tomato that dates from homesteading days. In fact, the original Homestead was a commercial variety developed jointly by the universities of Florida and South Carolina in 1952, and named for the Agricultural Experiment Station in the town of Homestead, Florida. Its current version, Homestead 24, was bred to provide hot-climate commercial growers in Florida, Texas, and Mexico with a Rutgers-type tomato—one that is fleshy and thick-walled, yet uniform in size and shape. Rutgers types are also fairly firm, but because Homestead 24 is less firm than most, it has gained wide acceptance as a home garden tomato in the South. It has a pleasantly sharp taste, and makes an excellent choice for home canning. Its 8-ounce fruits have small seed cavities, no core, and ripen evenly to the stem.

The determinate vines produce heavy crops when grown unstaked and left unpruned. Days to maturity: 82. Catalogues 2, 3, 15, 28, 33.

Husk tomato

Midseason

Standard

Determinate

Physalis pruinosa—known commonly as the husk tomato, ground cherry, strawberry tomato, or poha berry—is a relative of the tomato and is often listed alongside it in seed catalogues. Its fruits are the size of cherry tomatoes, measuring ¾ inch across, but the resemblance to tomatoes ends there. Yellow in color, they are enclosed in a papery husk; their distinctive flavor is sweet and aromatic, and their seeds resemble those of the eggplant.

While husk tomatoes are sometimes eaten raw, they are more

often used in jams or pies. They can be dried like raisins, frozen, canned, or candied, and they make a particularly good dessert when combined with oranges, lemon, and ginger.

The husk tomato is native to North America (unlike the ordinary tomato, which originated in South America), and has been cultivated here for more than 250 years. It is grown in the same way as the tomato, but is not staked. Its vines spread only 12 to 30 inches, and produce heavily in midsummer. They stand up well to heat and drought.

Goldie is an improved strain of husk tomato that is available from Johnny's Selected Seeds of Albion, Maine. It should not be confused with the yellow cherry tomato called Goldie, sold through the catalogue of Park Seed of Greenwood, South Carolina. Days to maturity: 75. Catalogues 6, 7, 11, 12, 14, 16, 20, 24, 25, 31, 32, 38.

*** Jet Star (VF)

Midseason
Hybrid
Indeterminate

Jet Star is rated excellent in all respects by home gardeners, but most say they grow it for its superior taste. It is most popular in the Pacific Northwest and the Midwest. Gardeners in Ohio rate it as their favorite variety—a sterling recommendation, since Ohio has a long tradition of home tomato gardening and was the site of much of America's early tomato breeding. (The Columbus suburb of Reynoldsburg even bills itself as the "Birthplace of the Edible Tomato.")

Fruits of Jet Star weigh about 8 ounces, are fairly firm and meaty, and have good resistance to cracking. The leafy indeterminate vines perform best when caged or staked. Days to maturity: 72. Catalogues 2, 13, 14.

** Jubilee

Midseason
Standard
Indeterminate

Jubilee, sometimes called Golden Jubilee, was like no tomato ever seen when it was introduced in 1943. It was unique not because it was yellow, since yellow tomatoes have been known since sixteenth century Spaniards first found the fruits in Mexico, but because it had a tangerine color gene that gave it a lovely yellow-orange hue that was different from the pure yellow of earlier varieties. Its unusual color gave it novelty value, and it soon became associated with the particularly mild, mellow taste of Jubilee.

The idea that Jubilee's color had anything to do with its taste was a misconception, however. Tomatoes taste mild because they have a high proportion of flesh to seeds and gel, and a resulting high sugar content. In the case of Jubilee and subsequent yellow-orange varieties, the fruits' seed cavities are irregular rather than symmetrical, so that none are large. This makes for fleshiness, and it is this characteristic, rather than pigment, that gives the tomatoes their mellowness.

All yellow tomatoes have the reputation of being low acid or subacid; consequently, they are supposed to be good choices for people who cannot tolerate red tomatoes. Actually, yellows have just as much acid as reds—but the acid taste is masked by high sugars and aromatic oils. And it is possible that yellows taste milder to some people simply because they are expected to, since psychological factors undoubtedly play a part in the perception of flavor (page 28).

Another misconception is that Jubilee takes its color from a high betacarotene content, and is therefore more nutritious than other tomatoes (betacarotene, a pigment, is the source of provitamin A, the substance that is converted into vitamin A when ingested). In fact, yellow and yellow-orange tomatoes, with the exception of Caro-Red and Caro-Rich, take their color from other carotene pigments, and have no more betacarotene than red tomatoes.

Jubilee was developed by Burpee's and received an All America Award in 1943 because of its high quality and adaptability to all climate zones. Its large, smooth fruits are round, have no shoulders, and average about 8 ounces, although they sometimes grow larger. A typical slice measures 3½ inches across.

While gardeners who are fond of yellow tomatoes rate Jubilee as their favorite, they say its yields are sometimes disappointing and its disease resistance only fair. Days to maturity: 72. Catalogues 2, 4, 7, 12, 15, 21, 24, 25, 28, 29, 31, 38.

All yellows developed since Jubilee include the tangerine color gene that gives them a yellow-orange color:

**Golden Boy is the only yellow hybrid tomato, and produces higher yields than Jubilee. Fruits average about 8 ounces. Indeterminate. Days to maturity: 78. Catalogues 2, 6, 10, 12, 14, 19, 26, 29, 31, 34, 38.

**Sunray (F) is almost identical to Jubilee, but with the advantage of resistance to fusarium wilt. Days to maturity: 78. Catalogues 1, 2, 10, 13, 15, 20, 22, 25, 30, 34, 39.

* Longkeeper

Midseason to late

Standard

Determinate

North Americans expect to eat fresh tomatoes year-round. Yet, understandably, they complain angrily about the quality of winter tomatoes in the supermarkets. Why, they demand, must tomatoes be so tough and flavorless? At the root of the problem is one factor: perishability. *Lycopersicon esculentum* simply doesn't keep well. Since out-of-season tomatoes must be grown in warm climates and shipped long distances to cold ones, they must be picked hard and green in order to survive the trip. Torn from the vine long before their sugars and acids have had a chance to develop properly, they will never be as good as they might have been, even though they will gradually turn red and somewhat softer.

Plant breeders have worked for years to solve the problem. They aspire to create a full-flavored tomato that can be picked ripe, instead of green, yet will remain edible for a month or more, in time to get from grower to grocer's shelf to kitchen. In the meantime, Burpee's has offered the home gardener a variety they say can be picked ripe in late fall and will stay perfectly edible until late winter. It is called Longkeeper.

Ironically, Longkeeper is not the result of years of experiments by horticultural scientists, but the find of one Bonnie Tewart of Wayne, Ohio. The tomato appeared as a cross between two open-pollinated varieties in Mrs. Tewart's garden—varieties she keeps secret, but says are "surprisingly obvious." Tomatoes harvested from the plant, and from later selections, kept their firmness, flavor, and texture months after being picked ripe. (Mrs. Tewart, who owned a café, harvested tomatoes in September and served them on hamburgers in January and February.) Burpee's, after a few years of trial plantings, bought rights to the tomato, and Mrs. Tewart became one of the rare home gardeners to have a home-grown variety made available to the general public.

The problem is that most home gardeners find Longkeeper little better than the rock-hard, flavorless tomatoes found in supermarkets in January. Its appearance is unusual: even when ripe on the vine, the 4- to 5-ounce fruit is almost colorless on the outside, yet red on the inside. It has a thick outer wall and several seed cavities. Its texture is dry and mealy, uncomfortably similar to that of supermarket types, and its taste is fairly acid, though unusual. Some gardeners find the taste interesting—not unpleasant, by any means, but not that of a real tomato, either.

If picked in September, Longkeeper can last until late February without changing flavor or texture. (Seeds within a number of the fruits may begin to sprout after about three months, however.) Gardeners should pick only ripe, unblemished fruits—they are ripe when the blossom end starts to turn from pale green to pale yellow-orange—and store them separately and unwrapped at room temperature. Like any tomatoes, they should never be stored in the refrigerator, nor will they take kindly to a sunny window sill. A shelf in a cool room, in the 65–68° range, is best.

Many gardeners, trying to time their Longkeepers to come ripe as close as possible to first frost, wait too long to plant—and at first frost have only a useless crop of unripe fruits. Gardeners in areas with long growing seasons should plant Longkeeper only two or three weeks later than other varieties. Gardeners in short-season areas should plant with early and midseason types. (Growers in Washington State and Wisconsin report that their Longkeepers never ripen on the vine. Growers in Texas, who can harvest in October or even November, say the tomatoes ripen well and keep until April.)

Some gardeners will find that Longkeeper, while hardly any-one's ideal fresh tomato, is indeed better than the tomatoes offered in the supermarkets in winter. Others will think they are not worth growing. Until breeders perfect the not-so-perishable ripe tomato, home gardeners can experiment with Longkeeper and decide for themselves. Days to maturity: 78. Catalogues 4, 6, 24, 37, 38.

** **Marglobe** (F)

Midseason

Standard

Large determinate

Along with Rutgers, Marglobe was the most popular all-purpose tomato of the 1930s and 1940s. Developed in 1925 by the United States Department of Agriculture, it was firm enough for shipping. And, as a result, it became an important commercial tomato. But it also became a favorite of home gardeners and canners because of its beautiful appearance—a perfect globe—and a pleasant flavor that was more sweet than acid.

Marglobe, which takes its name from its parents, Marvel and Globe, is still one of the most widely grown tomatoes in home gardens; commercially, it has been superseded by newer varieties. Its 6-ounce fruits are borne in clusters of 3 to 6 and are protected by a good cover of leaves. They tend to decrease in size as the season progresses. Thick-walled, the fruits ripen evenly but sometimes are subject to radial cracking. The plants do well in most parts of the continent, but sometimes have poor yields in hotter climates. Days to maturity: 73. Catalogues 2, 3, 4, 5, 6, 7, 10, 14, 15, 16, 18, 22, 25, 28, 29, 31, 33, 34.

***Marglobe Supreme** (F) is a later development from Ferry-Morse, bred to have fruits that are more uniform in size. Days to maturity: 73. Catalogues 7, 26.

*** **Marmande** (V)

Midseason

Standard

Large determinate

The French import Marmande, also known as De Marmande, is described in catalogues as "the classic European tastemaker." It has gained a reputation as the gourmet's tomato, and, although it is not widely available, is sought out by serious cooks and chefs for its unequaled flavor and texture.

Like most European tomatoes, Marmande has small fruits; they weigh about 4 ounces and are rather flattened in shape. Newer strains offer tremendous yields and resistance to verticillium wilt. The determinate vines prefer warmer climates than most European imports, and produce well in most parts of the U.S. and Canada. They are best grown in cages. Days to maturity: 74. Catalogues 8, 35.

*** **Moreton Hybrid** (v)

Midseason

Hybrid

Indeterminate

Northeastern gardeners rate Moreton Hybrid as the best-tasting home garden tomato. It has high concentrations of sugars, acids, and aromatic oils that combine to give the fruits a rich, robust flavor.

Vines are vigorous and have good foliage cover, producing 6- to 8-ounce fruits in clusters. Richly colored, even-ripening, and exceptionally meaty, they are excellent for canning as well as for eating fresh.

Developed by the Harris seed company of Rochester, New York, Moreton Hybrid adapts well to all parts of North America except the hottest southern areas. Days to maturity: 70. Catalogues 13, 23.

** **New Yorker** (v)

Early

Standard

Determinate

This northeastern commercial variety is an improvement on the Fireball type—a tomato that produces its entire crop over the space of a few days. It is superior to Fireball, however, in yields, disease resistance, fruit color, and foliage cover.

New Yorker's mild-tasting fruits weigh about 5 ounces and are harvested for about ten days early in the season. Later fruits decrease dramatically in size. The determinate vines set fruit well in cool weather, and grow best when left unstaked and unpruned. Days to maturity: 60. Catalogues 4, 13, 30, 31, 32, 38.

* Oxheart

Late

Standard

Indeterminate

Oxheart's unique shape, roughly resembling a heart, makes it one of the most unusual tomatoes available for growing at home. In all other respects, it is similar to Ponderosa: it is pink, large, fleshy, mild, and has few seeds.

Introduced in 1925 by the Livingston Seed Company of Ohio, Oxheart first appeared as a sport, or mutant, in a patch of Ponderosas grown by one of Livingston's customers.

Fruits usually weigh 12 to 16 ounces, but frequently grow up to 2 pounds. Most gardeners report their yields are fair at best, and sometimes poor. The sparsely foliaged vines should be grown in a cage to protect the fruits from sunscald. Days to maturity: 86. Catalogues 2, 3, 6, 7, 9, 10, 15, 19, 24, 25, 28, 31, 38.

* Patio (F)

Midseason

Hybrid

Dwarf determinate

Patio is a modern version of the old tree-type tomatoes of which Dwarf Champion is the best-known example—a bushy, dwarf plant that supports itself on sturdy stems. Small enough to grow in a 12-inch pot, it is a common choice of urban dwellers whose outdoor space is limited to balconies and courtyards; they say it is an exceptionally efficient producer, yielding a good crop of medium-sized tomatoes on a plant that takes up little room.

While most container gardeners are pleased with Patio's heavy production and tidy growth habit, they are disappointed with its taste and texture. Catalogues describe the tomatoes as sweet, but gardeners call them bland, mealy, and tough-skinned. The fruits weigh about 4 ounces and measure 2 inches across— unusually large for a plant that grows only 24 to 30 inches tall. Although the vines were bred to be self-supporting, they often require a short stake once the fruits begin to mature. Days to maturity: 70. Catalogues 1, 2, 6, 14, 17, 20, 24, 25, 26, 28, 30, 32, 33, 34, 37, 38, 39.

Patio Prize (VFN) is a similar, slightly earlier hybrid with better-tasting, thinner-skinned fruits. It also has superior disease resistance. Days to maturity: 67. Catalogues 5, 26, 29, 32.

** Pixie

Early

Hybrid

Dwarf, Cherry

Pixie is another tree-type dwarf in the class of Dwarf Champion and Patio. Its tomatoes are smaller, however, measuring only 1¾ inches across, and are classified as large cherry types. Nevertheless, they are said by gardeners to have the full-bodied flavor of large tomatoes.

Introduced in 1971 by Burpee's, Pixie was bred to produce a heavy crop on a compact plant that grows only 14 to 18 inches tall. The plants are usually grown in containers, but are also well suited to small gardens or flower beds, where they take up little space. Caging the plants will help protect the fruits from sunscald, since Pixie's foliage is sparse. Days to maturity: 52. Catalogues 4, 20, 36.

*** Ponderosa

Late

Standard

Indeterminate

Ponderosa is a grand old name among tomato varieties. An old-fashioned beefsteak type, it has never been excelled as a slicing tomato. It is also many gardeners' standard for flavor—sweet and mild, yet bursting with what can only be described as "real tomato taste."

Although the tomatoes are sometimes called Ponderosa Pinks, their flesh is actually red; translucent skin makes them appear pink from the outside. They usually weigh 9 to 16 ounces, but often grow to 2 pounds or more. One slice measures about 4 inches across, and shows few seeds and little gel. The flesh is almost solid, and has an exceptionally fine texture.

Like most beefsteaks, Ponderosa has a large core, and the fruits often grow in irregular shapes. They also have prominent shoulders and blossom end scars. But these imperfections, ironed out in later beefsteaks by modern plant breeders, are minor considerations for the gardeners who think Ponderosa's taste and succulence have never been equaled by the more streamlined varieties.

Ponderosa dates from 1891, when the Peter Henderson Seed Company put it on the market under the name of Henderson 400. That year, the company sponsored a competition among home gardeners to name the new variety, and Ponderosa was selected as the winner. The original variety was pink, but red and yellow types were also developed; they still are offered by some seed companies.

Yields of Ponderosa are not as large as those of new hybrids, but the strong indeterminate vines sometimes produce surprisingly well. Because they are rather sparsely foliaged, they are best grown in a cage to help protect the fruits from sunscald. Days to maturity: 90. Catalogues 2, 6, 7, 10, 13, 14, 15, 16, 24, 26, 28, 31, 32, 33, 34 (as Ponderosa Pink); 20, 29 (as Ponderosa Red).

*** **Porter**

Midseason

Standard

Indeterminate, Cherry

Texas's very own tomato was developed there in 1927 by seedsman V. O. Porter, who recognized the need for a tomato that would set fruit in hot weather and continue to bear all season long. Sometimes known as Little Porter, the tomato is small enough to qualify as a cherry type, but is egg-shaped and pink. It produces well through hot Texas summers, and for this reason remains one of the most commonly grown tomatoes in the state. It also is a favorite in other parts of the Southwest because of its reliably high yields and excellent taste.

Fruits weigh about 1 ounce, and are described in Porter's catalogue as about the size and shape of a pullet egg. They are smooth, coreless, ripen evenly, and are crack-resistant. Many gardeners like to use the meaty fruits for canning, preserves, or juice, as well as for eating fresh.

Indeterminate vines provide good foliage cover to protect the fruits from sunscald. Days to maturity: 70. Catalogue 28.

***Porter's Pride,** also known as Improved Porter, is quite different from the original Porter. It is not a cherry type, but is larger (3 ounces), rounder and bright red. It has all of the good characteristics of Porter, and is said to have the best keeping quality of any tomato. Its only flaw is a susceptibility to blossom end rot. Days to maturity: 70. Standard, indeterminate. Catalogue 28.

*** Red Cherry

Midseason

Standard

Indeterminate

Unlike large-fruited tomatoes, cherry types have managed to keep their quality when grown by large-scale commercial growers. Whether bought at the supermarket or grown at home, they are usually flavorful and succulent, and for this reason have become more popular as salad tomatoes than their large store-bought counterparts.

Gardeners have several varieties from which to choose (page 112). Some, such as Tiny Tim and Minibel, are intended primarily for use as ornamentals, and are bred to be compact enough for small pots or hanging baskets. Others, such as Sweet 100, grow on vigorous indeterminate vines that need stakes or cages to support their heavy, profuse clusters of fruits.

Just as the term "beefsteak" is used to describe tomatoes that are large and meaty, "cherry" is used generically to describe tomatoes that are small and round. But, just as Beefsteak is also the name of a specific variety, so is Red Cherry, which dates from the mid-nineteenth century in American seed catalogues. Its fruits measure ⅞ inch in diameter and weigh about ⅓ ounce. They have two cells filled with pulp and small seeds. The flavor is pleasantly mild and surprisingly sweet, but Red Cherry and most other cherry types lack the balance of acidity that most people consider essential for full tomato flavor. The fruits are borne in clusters of 7 to 9 on large, open-growing vines. Days to maturity: 72. Catalogues 1, 4, 6, 7, 13, 15, 19, 22, 25, 28, 32, 33, 38.

***Large Red Cherry** has fruits that measure 1½ inches in diameter. Days to maturity: 75. Catalogues 2, 5, 7, 13, 14, 20, 29.

***Yellow Cherry** is less widely available. It is identical to Red Cherry in every way except color. Days to maturity: 72. Catalogues 19, 32.

*** **Roma** (VF)

Midseason

Standard

Determinate, Plum

Roma is the best-known paste tomato, and still is used as the standard for flavor by home gardeners. Originally developed by the United States Department of Agriculture in 1955, it was improved in the early 1960s, by the Harris seed company of Rochester, New York, to resist verticillium and fusarium wilts. It makes a rich paste or sauce, and although many new commercial paste varieties, such as the Chicos, cook down to greater thickness, Roma tastes better. It does not need the addition of lemon juice or other seasonings to provide full-bodied flavor.

While Roma can be eaten fresh, its solid, thick flesh is dry and mealy in comparison to other tomatoes. Flavor and color are more intense, however, and the fruits can be peeled quite easily. They are used not only for paste, sauce, and catsup, but also for canning whole. Gardeners who make tomato juice often add a few Romas to the juice of larger varieties to provide extra body.

Romas are plum-shaped with a slight neck, and measure 3 by 2½ inches. Their compact determinate vines have good leaf cover and produce more tomatoes when grown on the ground or in cages; staking and pruning reduces yields. Days to maturity: 75. Catalogues 1, 2, 3, 4, 5, 6, 7, 9, 10, 13, 14, 15, 16, 20, 21, 22, 24, 25, 26, 29, 31, 32, 34.

**Romulus (vfn), imported from Holland, is resistant to nematodes as well as to verticillium and fusarium wilts. Days to maturity: 73. Catalogue 35.

**San Marzano, also known as Italian Canner, is the original Italian paste variety upon which Roma is an improvement. It is later and has drier fruits, but cooks down to superbly flavored paste. A large-fruited version of San Marzano, developed by Ferry-Morse, is the one generally offered in seed catalogues and at nurseries. It measures 3½ by 1½ inches, and is more rectangular in shape than Roma. Indeterminate. Days to maturity: 80. Catalogues 3, 4, 14, 16, 24, 25, 28, 29, 31, 32.

** Rutgers (f)

Late

Standard

Large determinate

Rutgers is one of the best known of the older tomato varieties, although it is not so old as most people think. Originally developed by the Campbell Soup Company in 1928 as a cross between Marglobe and a long-since forgotten variety called J.T.D., it was later fixed in type and named by the New Jersey Experiment Station, and was not widely grown until the 1940s. And—another surprise, considering its popularity as a home garden variety for eating fresh—it was intended for use as a commercial canning tomato. It offered something new: well-shaped, medium-sized tomatoes with fleshy, firm interiors and thick walls that did not split or crack when trucked from field to canning factory. (Other tomatoes, such as the beefsteaks, might have been just as meaty, but lacked the even ripening, firmness, smooth shape, and rich color of Rutgers.) Rutgers offered a new prototype for the tomato, and later varieties that share its qualities are frequently described as "Rutgers types."

By the 1950s, Rutgers had become more popular than Marglobe as an all-purpose tomato. It had higher yields, better disease resistance and superior flavor—a rich, mellow, full-bodied taste with little acidity.

In recent years, improved strains have been selected for different parts of North America, so that the range of tomatoes under the Rutgers name is fairly wide. As a rule, however, their fruits weigh about 7 ounces, and most are resistant to the concentric growth cracks that often appeared at the stem end of the original strain. The fruits are slightly flatter in shape than those of Marglobe, and ripen evenly from the inside out. They are still among the most popular varieties for home canning or for eating fresh.

The determinate vines produce their best crops when left unstaked and unpruned. They have good resistance to fusarium wilt. Days to maturity: 80. Catalogues 1, 2, 3, 4, 6, 7, 9, 10, 14, 15, 16, 21, 22, 25, 28, 31, 33, 34.

**Rutgers California Supreme is earlier and has slightly larger fruits, averaging about 8 ounces. Its name notwithstanding, it performs best in southern gardens. Days to maturity: 73. Catalogue 26.

***Rutgers Hybrid (VF) is a high-yielding hybrid with good disease resistance. Although widely adapted, it grows best in the northern and central states. Days to maturity: 80. Catalogues 2, 7.

***Rutgers 39 (VF) is an improved version with resistance to verticillium wilt as well as fusarium. Its vines are more compact and produce a heavier yield. Fruits are even less subject to concentric cracking. Days to maturity: 82. Catalogue 13.

** Small Fry (VFN)

Early

Hybrid

Determinate, Cherry

As cherry tomatoes go, Small Fry is said by gardeners to be blander than most. It rates high, however, for its reliability: it always produces a large crop, and it is sure to set fruit under adverse conditions. Its adaptability gained it an All America Award in 1970.

One-inch fruits are borne in clusters of 7 to 8 on 40-inch vines.

The plants grow well in 5-gallon containers, but should be staked or trellised. They also perform well in the open garden, especially when grown in a short wire cage. Days to maturity: 65. Catalogues 2, 3, 4, 6, 10, 13, 14, 15, 17, 20, 21, 25, 26, 28, 29, 30, 31, 32, 38.

** Spring Giant (VFN)

Early

Hybrid

Determinate

Although Spring Giant is an early variety, its fruits are as large as those of midseason types. Meaty and solid, they weigh 7 to 8 ounces. The plant's entire production is harvested over a relatively short time; this, and the fact the tomatoes have small cores, makes them a good choice for home canners.

Spring Giant grows well in all parts of the continent, and won an All America Award in 1967 for its good quality and adaptability. Southern gardeners report that it is one of the best earlies for their climate zone; they plant it so they can harvest a quick crop before their midseason and late varieties start maturing.

Like most hybrids, Spring Giant has high yields. Its determinate vines do not need staking. Days to maturity: 66. Catalogues 9, 19, 21, 24, 25, 28, 30, 31, 34, 36, 37.

** Springset (VF)

Early to midseason

Hybrid

Determinate

This high-yielding hybrid was developed for home gardeners and commercial growers in the northern states and Canada. A determinate, it produces most of its crop within the space of a few days. Fruits weigh 5 to 6 ounces, are fairly firm, and have good crack resistance. They are subject to sunscald, however, since the vines are sparsely foliaged. The problem can be overcome by growing the plants in cages to crowd the leaves together. Days to maturity: 65. Catalogues 6, 13, 14, 17, 25, 30, 32, 36, 38.

77

* Sub-Arctic Plenty

Early

Standard

Determinate

The four Sub-Arctic varieties were bred as extra-early tomatoes for extremely cold climates. The highest-yielding of them, Sub-Arctic Plenty, is also the most widely grown. Like the others, it produces fruits that are only slightly larger than cherry-size, averaging almost 2 inches in diameter.

The determinate Sub-Arctics produce most of their fruits within the space of a week or so. Their taste is described by gardeners as only fair, but still superior to that of many of the commercial varieties found in supermarkets.

The plants should not be staked or pruned, and can be placed as close together as 12 to 18 inches. Seed can also be sown directly in the garden, although the added growing time will increase the plants' risk of succumbing to wilt diseases. All of the Sub-Arctics are particularly susceptible to early blight. Days to maturity: 58. Catalogues 11, 12, 18, 27, 36, 37.

*Sub-Arctic Early is the earliest of the four varieties, but its fruits are smaller, averaging about 1⅜ inches. Days to maturity: 53. Catalogue 11.

**Sub-Arctic Maxi is the largest of the four, weighing about 2½ ounces. Flavor is slightly better than that of the other Sub-Arctics, and yields are almost as high as those of Sub-Arctic Plenty. Days to maturity: 62. Catalogues 18, 32, 36, 38.

*Sub-Arctic Midi is notable for the uniformity of its fruits. All are nicely rounded and measure about 2½ inches. Days to maturity: 56. Not sold through listed catalogues; available at nurseries in the northern states and Canada.

** Super Sioux

Midseason

Standard

Large determinate

Super Sioux, developed in Nebraska, was bred to set fruit during hot spells in the relatively dry climates west of the Mississippi. Although heat tolerant, it is not recommended for the humid southern states, where it is subject to fungus diseases.

Because the tomatoes ripen uniformly and are coreless, they make excellent canners. Fruits weigh about 6 ounces, and have a pleasantly acid flavor. Growing the large, open vines in a cage will help protect the fruits from sunscald. Days to maturity: 70. Catalogues 2, 5, 7, 28, 29.

*** Supersteak (VFN)

Midseason to late

Hybrid

Indeterminate

Supersteak, introduced by Burpee's in 1980, is modern plant breeding's version of the classic beefsteak tomato, with the imperfections taken out. We can be thankful that it has the same richness of flavor, but with slightly more tang, an attribute that pleases those who find beefsteaks too bland. Fruits are smoother than those of the old-fashioned types, and less subject to catfacing. They also have a smaller core and blossom end scar. The only questionable qualities the breeders failed to iron out of the fruits are green shoulders and a susceptibility to cracking.

Supersteaks average about 1 pound each, and often grow to 2 pounds. A typical slice measures 4½ inches across. Yields are better than those of older beefsteaks because of superior disease resistance—to verticillium wilt, fusarium wilt, and nematodes.

The vigorous plants should be supported on stakes or in cages, and usually do well in mild climates. In very hot areas, however, the plants may not set fruit; in cold climates, the fruits may not have enough time to ripen. Days to maturity: 80. Catalogues 4, 38.

*** Sweet 100

Early

Hybrid

Indeterminate, Cherry

Sweet 100 is undoubedly the most popular cherry tomato among home gardeners today. A hybrid introduced in the 1970s, it takes its name from its long, multiple-branched clusters, each of which may contain as many as 100 1-inch fruits. The fruits themselves are probably the sweetest of all tomatoes, yet are especially high in ascorbic acid (vitamin C).

Bred with the original South American wild tomato as a parent, Sweet 100 is among the highest yielding of the hybrids. Production can be increased even more by staking and pruning the vigorous indeterminate vines. The variety is adaptable to all climate zones, and is offered in catalogues and in nurseries from Canada to the South and Southwest. Days to maturity: 66. Catalogues 2, 3, 4, 6, 9, 11, 12, 13, 14, 15, 19, 20, 21, 22, 23, 26, 28, 29, 30, 32, 33, 34, 35, 36, 37, 38.

* Tiny Tim

Early

Standard

Dwarf ornamental, Cherry

Tiny Tim has the smallest tomatoes of the cherry types, with fruits measuring only ¾ inches across. Bright red and borne on a compact 15-inch vine, they are highly decorative, but are best grown strictly as an ornamental: gardeners rate them the least flavorful of the cherry tomatoes.

The miniature plants grow in pots or hanging baskets as small as 6 inches. If placed in a south-facing window, they will sometimes perform well indoors, but need several hours of direct sunlight each day to produce a good yield of fruits. Days to maturity: 55. Catalogues 3, 4, 6, 14, 18, 19, 21, 24, 25, 26, 28, 29, 31, 32, 35, 36, 37, 38, 39, 40.

Tree Tomato

A familiar full-page magazine advertisement asks, "Does 60 lbs. of Tomatoes from one yield sound incredible? Not if you own the amazing Tree Tomato." Even more incredible would be a yield of any tomatoes at all, since the plant in question is not a tomato, but *Cyphomandra betacea,* commonly known in other parts of the world as the tamarillo.

The term "tree tomato" was used until about forty years ago to describe self-supporting tomato plants—those with stout central stems that hold a bushy crown well off the ground until loaded down with mature fruits. An old pink variety, Dwarf Champion, and a modern hybrid, Patio, are the best-known examples. Gradually, however, "tree tomato" took on a different meaning in the minds of gardeners, and came to describe the most vigorous indeterminate types, including Climbing Trip-L Crop and Giant Tree, which can grow taller than a house if trained on a trellis.

Today, "tree tomato" is being used in advertisements to inaccurately describe the woody, huge-leaved plant that yields tart-tasting tamarillos. The tamarillo is, in fact, a member of the nightshade family, *Solanaceae*, to which tomatoes belong—but so, too, are potatoes, eggplants, peppers, and tobacco; and just as no sane gardener or cook would substitute one for the other, few would find the hard, acidic tamarillo an acceptable substitute for the tomato.

"Tree tomato" vendors claim in their advertisements that the gardener can "surprise . . . family and friends with ripe, just-picked tomatoes even during the coldest months of the year." Indeed, all concerned will be surprised: surprised that the perennial plant, which grows to 5 or 6 feet, may take several years even to begin producing fruits, and surprised even more when the fruits turn out not to be tomatoes, but tamarillos.

* **Walter** (F_1F_2)

Midseason

Standard

Determinate

Walter is a classic supermarket tomato, bred to take the bruises of the marketplace. Developed in the 1960s at the University of Florida, it is gradually being replaced by newer varieties, but still is used as a parent in the breeding of commercial tomatoes intended to ripen after being picked green.

Nevertheless, Walter is available to home gardeners. When picked ripe, the hard, 7-ounce tomatoes will taste much better than the Walters bought at the supermarket. But they still will be far inferior to varieties that were never meant for marketing.

Walter resists gray leaf spot, blossom end rot, and catfacing, as well as the two most common races of fusarium wilt. Days to maturity: 75. Catalogues 28, 34.

* White Beauty

Late

Standard

Indeterminate

Whites are thought by most people to be the least acidic of all tomatoes, on the assumption that the paler a tomato, the milder. In fact, pigment has nothing to do with acidity, and whites—like yellows—have just as high an acid content as reds. They do indeed taste milder, but only because their high sugar content masks acidity. And high sugar content is a consequence of exceptional meatiness, not of pigment.

Because whites are no less acid, their novelty value becomes the only reason to grow them. Their color, which is actually ivory or very pale yellow, makes them useful mainly as a conversation piece.

Fruits average about 8 ounces, but sometimes grow larger. They are fleshy and sweet, with few seeds. The indeterminate vines are adaptable to all parts of North America, but the slow maturation of the fruits makes them a risky choice in areas with short growing seasons. Plants should be grown in a cage or on a trellis. Days to maturity: 84. Catalogues 3, 11, 12 (as White Beauty); 10 (as Snowball).

A second white variety is available:

*White Wonder is slightly smaller than White Beauty, and is said by gardeners to be highly subject to catfacing. Days to maturity: 85. Catalogue 19.

*** Whopper, Park's (VFNT)

Midseason

Hybrid

Indeterminate

This home garden hybrid, developed by Park Seed of Greenwood, South Carolina, is thought of as a southern variety, but also grows well in other parts of the continent. It is consistently rated a favorite by gardeners, who praise its abundant production, large, full-flavored fruits, and excellent disease resistance.

Fruits weigh about 12 ounces and measure 4 inches across. They are borne in clusters and ripen evenly, without mottling, from blossom end to stem. They resist cracking and blossom end rot, as well as the most common tomato problems—verticillium and fusarium wilts, nematodes, and tobacco mosaic virus.

The strong-growing indeterminate vines are best grown in a cage, and offer a good leaf cover to protect the fruits from sunscald. Days to maturity: 75. Catalogue 26.

* **Wonder Boy** (VFN)

Late

Hybrid

Indeterminate

Introduced in 1959, when hybrids were still something of a novelty, Wonder Boy was notable for its high yields and its resistance to verticillium and fusarium wilts. The version that is available today also has resistance to nematodes. Gardeners in all but the coldest areas can be assured of big crops, but most have the same complaint: Wonder Boy's flavor is inferior to that of most of the older open-pollinated varieties and many of the newer hybrids. While the fruits are particularly meaty, they have little taste.

Because Wonder Boy sets fruit well in hot weather and provides a good leaf cover to shield fruits from sunscald, it was widely grown in the Southwest in the 1960s and early 1970s; today it has largely been superseded by newer and better-tasting varieties. Fruits are large, weighing up to 1 pound, and are described by some catalogues as being more uniform in size than Better Boy. The vigorous, tall-growing vines should be staked or caged. Days to maturity: 80. Catalogues 5, 6, 7, 10, 17, 22, 25, 31, 32, 34.

** Yellow Pear

Midseason

Standard

Indeterminate

Yellow Pear is one of the oldest varieties that is still widely planted. It was first described in horticultural literature in the United States in 1848, and has been generally listed by seedsmen since the late 1800s. Gardeners today grow it as an ornamental, or use its sweet fruits for preserves or pies. The 1¾- to 2-inch long, pear-shaped tomatoes, clear yellow with a waxy finish, also make a novel hors d'oeuvres when served on relish trays or with a dip.

Fruits are borne in clusters on luxuriant plants that are adaptable to all climate zones. Days to maturity: 76. Catalogues 2, 3, 4, 7, 10, 12, 13, 14, 16, 19, 20, 21, 22, 24, 25, 26, 28, 29, 32, 38.

Red Pear matures earlier than Yellow Pear but otherwise is identical except for its color. Because it lacks the novelty value of the yellow variety, it is less commonly grown. Days to maturity: 70. Catalogues 3, 10, 11, 12, 13, 14, 15, 19, 24, 26, 32, 38.

** Yellow Plum

Midseason

Standard

Indeterminate

Although plum tomatoes are generally used for making tomato paste or sauce, the old varieties called Yellow Plum and Red Plum, which date from the 1860s, are more often used for canning or preserves. (Newer plum varieties make better paste and sauce because their solids cook down to greater viscosity.)

Yellow Plum's fruits resemble those of Yellow Pear, but without the distinct neck. They are oval or plum-shaped, usually grow to 2 inches long and 1½ inches thick, and weigh about 1 ounc Their flavor is mild and sweet, and they are sometimes described as subacid, although their acid content is just as high as that of red tomatoes.

Many mail-order seed companies sell a seed mixture of small-fruited varieties, including Yellow Plum, Red Plum, Red Pear, and Yellow Pear; they are attractive when canned together in glass jars, and also combine to make unusual preserves. Days to maturity: 70. Catalogues 7, 13, 14, 19, 24, 25, 26, 29, 38.

Red Plum is identical to Yellow Plum except for its color. Days to maturity: 70. Catalogues 14, 19, 26, 32, 38.

225 ADDITIONAL VARIETIES BRIEFLY DESCRIBED

EARLY VARIETIES

Alfresco   60 days. A Canadian specialty, notable for its exceptionally heavy yields. Firm fruits are uniform in size, weighing about 4 ounces and measuring 2 inches across. Determinate plants are compact enough to make Alfresco a fine choice for short-season growers with small gardens. Hybrid. Catalogue 38.

Basket King   55 days. Page 40.

Benewah   60 days. Developed in Idaho to set fruit in cool summers, and particularly suited to mountain gardens in the Pacific Northwest. Fruits are larger than those of most early types, weighing up to 7 ounces and measuring almost 3 inches across. Taste is slightly fruity. The determinate vine needs no staking, and yields are generally good. Standard. Catalogue 24.

Big Early, Burpee's   62 days. This indeterminate hybrid produces thick-walled fruits that compare favorably in size with main crop tomatoes, averaging 7½ ounces throughout the growing season. Fruits ripen evenly to an intense red. Big Early is said to be adaptable to all climates, but gardeners in the Southeast rate its yield as only fair, disease resistance as poor. The vigorous vine is best grown in a cage or on a trellis. Burpee's offers a similar, improved hybrid called Early Pick. Catalogues 2, 4, 22, 29.

Bitsy (VF)   52 days. Bitsy produces small tomatoes, though not so miniature as the name suggests. Sweet, red fruits are larger than those of cherry types, measuring 2 inches across and weighing about 3½ ounces. The determinate plants are especially recommended for container growing. Hybrid. Catalogue 26.

Bonner 60 days. This large cherry type was bred for mountain gardens in Idaho and the Pacific Northwest, where the growing season lasts only from early July to late September. Fruits, weighing 2 ounces and measuring 1½ inches across, ripen all at once on the compact determinate vine. Gardeners in areas with unusually cool summers say Bonner never fails to set fruit. Standard. Catalogue 11.

Break o' Day 63 days. This bright orange-red tomato has been grown by home gardeners and commercial growers since the 1930s. Its main recommendation is size: fruits are larger than those of most early types, weighing 6 to 7 ounces and measuring 3 inches across. Yields are good, but fruits are subject to sunscald because of the sparse foliage of the indeterminate vine. Standard. Catalogue 2.

Burpeeana Early 58 days. One of the higher-yielding early tomatoes in the East and Midwest. Mild-flavored fruits are medium in size, averaging 5 ounces, and are borne in clusters of 5 to 8 on an indeterminate vine. This variety is frequently grown in greenhouses by commercial growers. Hybrid. Catalogues 4, 36, 37.

Bush Beefsteak 62 days. Page 42.

Cabot 58 days. Canadian commercial variety developed to have medium-sized, extra-firm fruits that ship well. Standard, determinate. Catalogue 40.

Cherry Grande (VF) 58 days. As the name suggests, these cherry tomatoes are relatively large, measuring 1½ inches across. Gardeners in all parts of the country recommend the variety for its heavy yield of sweet fruits. The determinate plants provide good foliage cover to prevent sunscald. Hybrid. Catalogues 34, 38.

City Best (VFN) 60 days. Also known as CB, this compact Southern specialty is ideal for urban gardens with limited space, and for container growing on patios or balconies. Medium-sized fruits, about 5 ounces each, are borne on determinate plants. Hybrid. Not sold through listed catalogues; available at nurseries in the South.

Doublerich 60 days. Page 53.

Earliana 62 days. Page 56.

Earlibright 60 days. Developed in Canada as a commercial variety, Earlibright is frequently grown by home gardeners in northern areas because of its ability to set fruit in cool summer weather. Fruits average about 4 ounces and have a slightly sweet taste. Determinate plants are compact. Standard. Catalogues 18, 32.

Earlirouge 63 days. This Canadian commercial variety produces sweet fruits that are larger and more uniform in size than those of most early varieties. They average about 6 ounces. Because it sets fruit well at extremes of temperature,

Earlirouge is grown in the southwestern U.S. as well as in the northern states and Canada. Standard, determinate. Catalogues 18, 27, 32, 38.

Early Big (vf)   65 days. A commercial variety—not to be confused with Burpee's Big Early—that performs well from California to the East, producing heavy crops of firm, 7-ounce fruits. Hybrid, determinate. Not sold through listed catalogues; available at nurseries.

Early Bird   57 days. This commercial variety was developed specifically to withstand the abuse of transplanting on a mass scale. Its hard fruits make it suitable for shipping, but a poor choice for home gardeners. Fruits average 5½ ounces each. Standard. Not sold through listed catalogues; available at nurseries in the northern states and Canada.

Early Cascade   60 days. Page 57.

Early Chatham   55 days. This old Canadian variety is eternally popular because of its excellent flavor. Determinate plants, able to stand colder-than-usual night temperatures, produce firm, 6-ounce fruits. Standard. Catalogue 36.

Early Cherry   56 days. One of the earliest of the cherry types, with a heavy yield of sweet, 1-inch fruits. The determinate vine is compact; like most cherry tomatoes, it is a good choice for container growing. Standard. Catalogue 1.

Early Giant   65 days. This is promoted as one of the larger early tomatoes, with fruits in the ½-pound range. It grows best in warmer regions, setting fruits even in hot southwestern summers. Gardeners in the far north criticize the variety as "neither early, nor giant." Indeterminate vines produce heavy yields of smooth, thick-walled fruits. Hybrid. Not sold through listed catalogues; available at nurseries.

Early Girl   58 days. Page 57.

Early Pick, Burpee's (vf)   62 days. Developed by Burpee's as an improvement on their Big Early variety, this indeterminate plant sets fruit in a wider range of temperatures and also has more disease resistance. Burpee's especially recommends it for the West Coast, where low night temperatures can impair the fruit-setting ability of many varieties. Fruits average about 7 ounces, but often grow up to 1 pound each. They are solid and juicy, with rather mild flavor. Globe-shaped fruits, slightly flattened, ripen evenly so that there are no green shoulders. Hybrid. Catalogue 4.

Early Red Chief   65 days. A northern commercial variety. Determinate vine bears heavy crops of 10-ounce tomatoes over a long growing season. Used mainly for fresh market sales and commercial canning. Standard. (Not to be confused with the midseason hybrid called Red Chief, a southern home garden variety developed by the H. G. Hastings company.) Not sold thorugh listed catalogues; available at nurseries.

Early Salad, Burgess   45 days. This large cherry type is perhaps the earliest of all tomatoes. Compact determinate plants, 6 to 8 inches high and spreading 2 feet, produce up to 300 sweet tomatoes about 1½ inches across. Fruits often remain on the plant for a full month, even in hot weather, without losing quality. Hybrid. Catalogues 3, 26.

Early Temptation 46 days. This variety is commonly grown in eastern Canada as the earliest non-cherry type. Fruits are fairly small, averaging about 3 ounces, but ripen evenly and have good flavor. The dwarf determinate plant requires no staking or pruning. Hybrid. Catalogue 40.

Early Wonder, Burgess 60 days. Fruits of this variety are enormous compared to those of other early tomatoes: sufficient watering and fertilizing will produce meaty, mild-flavored tomatoes weighing up to 14 ounces and measuring almost 4 inches across. Despite their high yield of large fruits, the determinate vines are small enough to be planted only 2½ feet apart. Standard. Catalogue 3.

Easy Peel 65 days. One of the few tomato varieties bred to have skin that slips off easily without scalding. Bright red fruits weigh 5 to 6 ounces. The obvious recommendation of this variety is for home canners who want to simplify their task. Hybrid. Catalogue 26.

Extra Early, Park's (VFNT) 52 days. The fruits of Extra Early are unusually large for a variety with such a short maturation period: they average 4 to 5 ounces. The variety is popular in the South, but is rated higher by gardeners for its disease resistance than for its taste or yield. Indeterminate vines should be staked and pruned to produce full-sized fruits. Hybrid. Catalogue 26.

Fireball 60 days. Page 59.

Florida Basket 56 days. This cherry type, similar to Florida Petite, is ideal for hanging baskets. Thin-skinned, sweet fruits are larger, measuring 2 to 2½ inches across. Standard, compact determinate. Catalogue 32.

Florida Petite 60 days. Page 61.

Gardener (VF) 63 days. Indeterminate hybrid developed for northeastern commercial growers and home gardeners from Pennsylvania through the easternmost parts of Canada. Firm fruits weigh about 6 ounces and resist cracking. The plants should be grown on stakes, trellises, or in cages. Not sold through listed catalogues; available at nurseries in the northeast.

Gem State 58 days. This cherry type produces heavy yields of 1½ to 2 ounce fruits on thick-stemmed determinate plants. Ideal for small gardens or containers. Standard. Catalogue 18.

Golden Delight 60 days. One of the earliest yellow tomatoes. Compact determinate plants produce 5-ounce fruits that are meaty and crack resistant. Their flavor is mild, but, like all yellow tomatoes, they have an acid content as high as that of red varieties. Short maturation time makes this one of the more commonly grown yellows in Canada. Standard. Catalogues 12, 32, 40.

Golden Sunrise   58 days. This English yellow tomato is more strongly flavored than most yellows. Smooth fruits, weighing 6 to 7 ounces, are borne on indeterminate vines. Standard. Catalogue 35.

Hasty Boy (vf)   60 days. An early variety bred for southern gardens. Fruits are sweet, weigh about 6 ounces and are crack resistant. Hybrid, determinate. Catalogue 15.

Ida Gold   59 days. The earliest of all the yellow types, producing 2-ounce fruits on a bushy determinate plant. Standard. Catalogue 18.

Jetfire (v)   60 days. A Canadian commercial tomato, developed as an improvement on the Springset variety. Fruits average 7 to 8 ounces, are firmer than home garden types, and are said by some gardeners to be rather rough. Dwarf determinate vines do not need staking. Resistant to verticillium wilt, with tolerance for fusarium wilt. Hybrid. Not sold through listed catalogues; available at nurseries in Canada.

JSS #3570   60 days. The flavor of this Maine specialty is sweeter than that of most early varieties, including the popular Sub-Arctics. Fruits are quite small, however, weighing about 2 ounces each. Foliage is sparse and fruits are subject to cracking, but the plant is resistant to the blight that affects many early types. Indeterminate vines should be staked. Standard. Catalogue 18.

Juice, The (vf)   65 days. This processing tomato was bred to have an extremely high juice content. It is recommended, obviously, for home gardeners who want tomatoes for juice-making rather than for slicing. Fruits weigh 6 to 8 ounces. This variety grows best in the western half of the U.S. Hybrid, determinate. Catalogues 10, 11, 37.

Lark   59 days. Developed as a commercial variety for growers in North Dakota. Determinate vines produce hard, medium-sized tomatoes resistant to cracking and catfacing. Standard. Not sold through listed catalogues; available at nurseries in North Dakota.

MacPink   62 days. This pink variety is an introduction from the MacDonald campus of McGill University in Canada. It is the earliest of the pink tomatoes, and the only pink determinate variety available to home gardeners. Fruits are mild and sweet and weigh about 5 ounces each. Standard. Catalogue 38.

Manitoba   60 days. This Canadian variety, with 6-ounce fruits, is being superseded by newer and better-tasting varieties. Earliness is its main recommendation. Standard, determinate. Catalogues 32, 36, 37, 39.

Montfavet 63-5   64 days. This hybrid, from France, is grown for its superior taste and high yield. Medium-sized fruits grow on a vigorous indeterminate vine. Not recommended for hot climates. Catalogue 35.

Mustang   58 days. This is one of the older Canadian hybrids, long a favorite in Manitoba. Determinate vines produce high yields of small to medium-sized fruits. Mustang is said to be one of the best varieties for ripening indoors late in autumn. Catalogues 37, 39.

New Yorker   60 days. Page 69.

Orange Queen   65 days. Bright orange in color, and earlier than most yellow types. Fruits weigh 4 to 6 ounces and are slightly flattened. Catalogues describe this and other yellow varieties as "low acid," although yellows are actually as acidic as reds; their taste, however, is quite mild. Determinate, standard. Catalogue 32.

Perron 50   65 days. This Canadian commercial variety is similar to Springset, but is firmer and larger. A pleasant, sweet flavor makes it popular with home gardeners. Determinate vines produce slightly flattened 10-ounce fruits that measure almost 4 inches across. Hybrid. (Not to be confused with Peron, a South American import.) Catalogue 38.

Ping-Pong   55 days. Decorative compact plants produce cherry tomatoes the size of ping-pong balls, with sweet flavor and tender skin. Fruits are crack resistant and stay on the vine when ripe. A good choice for container growing. Determinate, standard. Catalogue 39.

Pink Pearl   50 days. One of the few pink cherry tomatoes. Tiny fruits, about ¾ inch across, are a distinctive rosy pink, and have sweet flavor. The compact determinate plant is slightly larger than that of the popular Tiny Tim variety. Decorative plant for sunny windows. Standard. Not sold through listed catalogues; available at nurseries.

Pixie   52 days. Page 71.

Plainsman   65 days. This variety was developed in Texas as a processing type, yet became popular as an early tomato among home gardeners from west Texas to Colorado because it sets fruit well in areas with high day temperatures and low night temperatures. It also has ample foliage cover to protect the fruits. Compact determinate plants produce firm 5- to 6-ounce tomatoes. Standard. Catalogues 5, 28.

Prairie Pride   54 days. A dwarf compact Canadian variety that produces 5-ounce, thick-walled tomatoes with good keeping quality. Popular in Canada as a low acid type. Determinate vines are self-supporting. Standard. Catalogue 39.

Presto   60 days. This small tomato, classified as a large cherry or small salad type, is said by home gardeners to have exceptionally good flavor. Determinate vines grow about 2 feet tall and, if grown in containers, should be staked or trellised. Hybrid. Catalogue 13.

Quebec #5   60 days. This Canadian variety with dark red, 5-ounce fruits does well in soils with a high nitrogen content. Standard, indeterminate. Catalogue 38.

Red Cushion   65 days. This hybrid has a high yield of typical cherry tomatoes. It is a good choice for growing in containers, since its determinate vine spreads no more than 18 inches. Has no resemblance to the old fashioned "cushion" tomatoes with deeply recessed centers, but is so named because the shapely, compact plant is virtually covered with red fruits. Catalogue 29.

Revolution (VF$_1$F$_2$)   62 days. A jointless variety with firm, 7-ounce fruits. Determinate plants are bred to be compact enough to grow close together, thus increasing yields for commercial growers. Standard. Catalogue 34.

Rocket   50 days. This Canadian variety, also known as Centennial Rocket, was developed for areas with extremely short growing seasons. Slightly firm fruits weigh 3 to 4 ounces. The small determinate vine is suitable for container growing. Standard. Catalogues 32, 36, 37, 39.

Roza (VF)   63 days. Two qualities make Roza unusual: it is one of the few varieties suited to desert climates, and it has much more vitamin C than most tomatoes. Determinate plants produce firm, medium-sized fruits that are satisfactory for eating fresh, excellent for canning. Resistant to curly top virus, as well as verticillium and fusarium wilts. Hybrid. Not sold through listed catalogues; available at nurseries in the West.

Salad Top   60 days. The tiny determinate plant of this variety—topping out at no more than 8 inches—can be grown in a 6-inch pot on a sunny window sill, patio, or balcony. Cherry tomatoes are 1 inch across. Standard. Not sold through listed catalogues; available at nurseries.

Scotia   55 days. Popular in Canada because of its ability to set fruit in unusually cool weather. Fruits weigh about 4 ounces and have green shoulders. Canadian gardeners describe Scotia as a dependable early cropper. Determinate, standard. Catalogues 32, 40.

Shumway's Sensation   55 days. A specialty of the R. H. Shumway Company of Illinois, this high-yielding hybrid produces smooth 6-ounce fruits, setting 5 to 7 to a cluster. Midwestern gardeners rate it the heaviest producer among the early varieties. Indeterminate vine offers exceptionally good foliage protection. Catalogue 31.

Sprinter (VF$_1$F$_2$)   58 days. The medium-sized fruits of this processing tomato are firmer and less flavorful than home garden types. The compact determinate plant has a high yield, however, and resists early blight as well as verticillium and fusarium wilts. Hybrid. Catalogue 34.

Starfire   56 days. A Canadian variety that grows best in sandy soil. Fruits average 6 ounces and are borne on compact determinate vines. Standard. Catalogues 32, 36, 37, 38, 39.

Starshot (V)   55 days. The compact, determinate growth habit and high yields of this variety make it a good choice for urban gardeners in the Northeast.

Grown on short stakes in containers, each plant yields about 25 3-ounce fruits. The plant is a vigorous grower, even in high smog areas. Standard. Catalogue 32.

Stokes Early 54 days. Recommended mainly for commercial growers who want fruits to maintain their size throughout the growing season. Fruits average 4 ounces. Indeterminate vines should be staked. Hybrid. Catalogue 32.

Stokesalaska 55 days. Mild, sweet flavor and thin skin make this a commonly grown variety in the northern U.S. and Canada. Bushy determinate plants spread about 18 inches and produce as many as 60 small fruits weighing almost 2 ounces each. Suitable for growing in containers. Standard. Catalogue 32.

Sub-Arctics 53–62 days. Page 78.

Sun-Up (F) 60 days. A commercial hybrid developed in Missouri. Fruits are similar to the Fireball variety, but are slightly larger. Compact determinate plant offers good foliage cover. Because fruit set is concentrated over a short period, the variety is a good choice for home canners. Not sold through listed catalogues; available at nurseries in Missouri.

Super 23 58 days. This high-yielding hybrid produces better-flavored fruits than those of most early commercial varieties, making it suitable for home gardeners. Fruits are large, weighing 8 to 10 ounces. Indeterminate. Catalogue 7.

Sweet-N-Early (VF) 55 days. This hybrid bears 7 to 9 sweet fruits, averaging 3 ounces, on each of its clusters. Indeterminate. Not sold through listed catalogues; available at nurseries.

Swift 54 days. A northern favorite with small, flavorful fruits measuring 2 inches across, Swift dependably sets fruit at low temperatures. The bushy plants do not need staking or pruning. Determinate, standard. Catalogue 32.

Tiny Tim 55 days. Page 80.

Ultra Girl (VFN) 56 days. A favorite of gardeners in Wisconsin and Minnesota because of its large fruits, disease resistance, good flavor and earliness. Firm red fruits weigh 7 to 9 ounces; they are crack resistant and do not have green shoulders. The large determinate vines are usually staked. Hybrid. Catalogue 32.

Vigor Boy (VF) 63 days. This exclusive from Gurney's Seed & Nursery produces 7- to 8-ounce fruits that ripen evenly with no green spots. Fruits are smooth-skinned and meaty, borne on determinate vines. Hybrid. Catalogue 12.

Wayahead, Improved 63 days. Jung Quality Seeds of Wisconsin developed this improved version of the old Wayahead. Smooth, solid fruits weigh 6 to 7 ounces and have good flavor. Determinate, standard. Jung also offers, at a higher price, seed selected from the best plants of this variety. Catalogue 19.

Whippersnapper   52 days. Dark pink cherry tomato with oval fruits measuring 1 inch across. The compact determinate plant holds as many as 100 fruits at once. Good for hanging baskets. Standard. Catalogue 18.

MIDSEASON AND LATE VARIETIES

Abraham Lincoln   80 days. Page 38.

Ace   85 days. Page 39.

Ace 55   80 days. Page 39.

Ace-Hy   76 days. Page 39.

Alicante   68 days. This is the classic English breakfast tomato, prized by gourmets. Fruits weigh about 4 ounces and are exceptionally meaty. Standard, indeterminate. Catalogue 8.

Americana   68 days. A high-yielding hybrid that produces smooth, firm 6- to 7-ounce fruits measuring 3½ inches across. Popular in northern states and Canada because of its ability to set fruit under adverse conditions. Large determinate. Catalogue 38.

Atkinson   75 days. Page 40.

Avalanche (F)   70 days. This Missouri-bred hybrid is noted for its extremely high yields. Fruits, averaging 6 to 7 ounces, have good resistance to cracking and ripen uniformly. Indeterminate vines should be staked and pruned to maintain fruit size. Catalogues 2, 7, 21.

Basket Pak   76 days. Typical midseason cherry tomato, developed by Burpee's to have high yields. Fruits measure 1½ inches across. Standard, indeterminate. Catalogue 4.

Basketvee   70 days. Because it ripens from the inside out, this large commercial variety has better taste than most commercial types. Fruits weigh 8 to 9 ounces and resist cracking and catfacing. Standard, determinate. Catalogue 32.

Beefmaster   80 days. Page 42.

Beefsteak   82 days. Page 41.

Bellarina   73 days. This paste variety is used by home gardeners for catsup, purees, sauces, and tomato paste. Small fruits—averaging 3½ ounces—are meaty, with little juice, and have a pleasantly mild flavor. Determinate vines produce 80 to 100 fruits per plant. Standard. Catalogue 12.

Bellstar   74 days. This Canadian plum tomato has fruits that weigh 4 to 6 ounces—nearly twice as large as most plum types. Standard, determinate. Catalogue 18.

Better Boy   72 days. Page 43.

Better Bush (vFN)   78 days. Introduced in 1984, this variety is hailed by seedsmen as a breakthrough: it bears large tomatoes on a compact indeterminate vine with short internodes. The bushy plant grows only 3 feet tall, yet produces as large a crop as a plant twice its size. Fruit weighs up to 8 ounces. Hybrid. Catalogue 26.

Big Boy   78 days. Page 44.

Big Girl (vF)   78 days. This is Burpee's effort to improve on their older Big Boy variety, with more disease resistance, no green shoulders, and less cracking. Yet most home gardeners still prefer Big Boy for its superior taste. Big Girl's indeterminate vines bear fruits weighing up to 1 pound. Resistant to verticillium and fusarium wilts, but prone to blossom end rot. Hybrid. Catalogues 2, 4.

Big Pick   70 days. Page 45.

Big Sandy   90 days. An heirloom variety thought to have originated in West Virginia. Huge, pale red fruits weigh up to 2½ pounds, measure 6 inches across and are almost coreless. Like all beefsteak types, they are meaty, with small seed cavities, and have a delicious, mild flavor. Old-timers speculate that the name derives from "big sandwich," since one slice will cover a slice of bread. Indeterminate vines should be staked and pruned. Standard. Not available commercially; seeds exchanged through seed savings organizations.

Big Set   65 days. Page 46.

Big Seven (vFN)   77 days. This indeterminate hybrid, which grows best in the eastern half of the U.S., is noted for its heavy yields. Fruits weigh up to 1 pound, but are rated lower in flavor than the similar Better Boy and Big Pick. Catalogue 34.

Blazer   68 days. Adaptable to a wide range of climates, as demonstrated by its sale in seed catalogues from regions as diverse as Canada and Texas. Good-flavored fruits weigh about 9 ounces and resist cracking. Large determinate vines can be staked or allowed to grow as a bush. Catalogues 28, 39.

Bonnie (vFN)   70 days. Developed for Bonnie Plant Farm in Alabama, this indeterminate hybrid is sometimes known as Bonnie Farms. Gardeners give it high marks for its heavy yield and excellent flavor. Fruits usually weigh 6 to 8 ounces, although Texas gardeners report the fruits sometimes do not come up to size. Not sold through listed catalogues; available at nurseries in the South and Midwest.

Bonny Best   70 days. Page 46.

Bonus   75 days. Page 47.

Bradley   74 days. Page 48.

Bragger 85 days. Page 49.

Brandywine 85 days. Page 49.

Burgis (vf₁f₂) 78 days. Developed in Florida as a higher yielding, larger-fruited version of a common supermarket variety, Flora-Dade. Like Flora-Dade, Burgis is jointless, which means it can be picked stem-free when mechanically harvested. Standard, determinate. Not sold through listed catalogues; available at nurseries in Florida.

Burnley Bounty 75 days. An Australian tomato that is the most popular of the nonhybrid types grown there. Indeterminate vines produce heavy yields of 5-ounce fruits with a pleasantly sharp flavor. Best adapted in the U.S. to southern California and Florida. Catalogue 35.

Burpee's VF 72 days. A deliciously mellow hybrid that is rated high by gardeners for its flavor and good performance. Meaty, 7- to 8-ounce fruits have smaller cores than many other varieties. As the name describes, indeterminate vines are resistant to verticillium and fusarium wilts; the fruits are crack resistant. Burpee's says this variety is a good choice for greenhouse growing. Catalogues 4, 7, 22.

Cal-Ace 80 days. Page 39.

Calypso (vf₁f₂) 83 days. A commercial variety finding favor among home gardeners in humid climates because of its excellent resistance to foliage diseases. Fruits, weighing 7 to 8 ounces, are firmer than most home garden types, and are well suited to canning or freezing. Determinate vines, with good foliage cover, are resistant not only to verticillium wilt and both races of fusarium wilt, but also to early blight and gray leaf spot. Standard. Catalogue 1.

Campbell varieties 70–78 days. Page 51.

Cannonball 71 days. Developed in North Dakota, this determinate variety is similar to the better known Spring Giant. Fruits weigh 6 to 7 ounces and are firm, although not so firm as the name suggests. Standard. Not sold through listed catalogues; available at nurseries in North Dakota.

Caribe (vf₁f₂) 68 days. Hard commercial variety adapted to humid areas. Fruits average 7 ounces and have good foliage protection on a compact determinate bush. Resistant to early blight and gray leaf spot as well as verticillium and fusarium wilts. Not sold through listed catalogues; available at nurseries.

Caro Red 78 days. Page 52.

Caro Rich 80 days. Page 53.

Celebrity 70 days. Page 53.

Champion (vfnt) 70 days. Introduced in the early 1980s, this hybrid has steadily gained favor for its disease resistance, wide adaptability, and good flavor. Indeterminate vines produce an exceptionally heavy yield of 8- to 9-ounce tomatoes throughout the growing season, with fruit-setting ability unaffected by hot weather. Called Big Red Champion in the catalogue of Henry Field Seed & Nursery of Shenandoah, Iowa. Catalogues 10, 24.

Chico III 75 days. Page 54.

Climbing Trip-L Crop 90 days. Page 55.

Cocktail 76 days. This 1-inch cherry tomato has been superseded in recent years by improved types such as Sweet 100. Bushy determinate plants are suitable for container growing. Standard. Not sold through listed catalogues; available at nurseries.

Coldset 70 days. This Canadian variety was bred to set fruit in cooler-than-usual temperatures. Its flavorful 5-ounce tomatoes have a deep red interior color. Standard, determinate. Catalogues 12, 32.

Colossal 90 days. A huge beefsteak type offered by Burgess in four colors—crimson, golden, red, or yellow. All four produce typically mild, meaty beefsteak fruits that average 14 ounces but often grow up to 2½ pounds and measure 6 inches across. Burgess also offers a higher-yielding hybrid version called Super Colossal, which matures about five days earlier. Its fruits—just as large as those of the original—are more uniform in size and shape, and more flavorful. Indeterminate. Catalogue 3.

Contessa (vF₁F₂) 68 days. Commercial variety grown from California to Florida. Hard, green-shouldered fruits weigh 7 ounces. Resistant to fusarium wilt, nematodes, early blight, and gray leaf spot. Not sold through listed catalogues; available at nurseries.

Crack-Proof, Burgess 80 days. Developed to resist cracking under the most adverse conditions. Medium-tall determinate vines produce large, slightly flattened fruits that weigh up to 9 ounces and ripen without green shoulders. A good choice for home canners, since the thin, elastic skin is easy to peel. Standard. Catalogue 3.

Creole 72 days. Developed in Louisiana for warm, humid climates, this indeterminate variety resists foliage diseases, blossom end rot, and fusarium wilt. Smooth, medium-sized fruits are slightly firm, but have good, strong tomato flavor and plenty of juice. The plant is best grown on a stake or in a cage. Standard. Not sold through listed catalogues; available at nurseries in Louisiana.

Crimson Trellis 78 days. An indeterminate, or staking, version of a Canadian bush type called High Crimson. Slightly flattened fruits weigh about 7 ounces. Used as a spring greenhouse crop in New England, and as a late variety in northern gardens. Standard. Catalogue 18.

Crimsonvee 72 days. A Canadian commercial variety used for tomato paste and catsup. Very firm fruits—almost square in shape—grow on bushy determinate plants and are resistant to cracking. Standard. Catalogue 32.

Delicious 78 days. Page 56.

Dixie Golden Giant 84 days. This huge yellow beefsteak, long a southern favorite, is no longer listed in catalogues. Fruits, with the mild, sweet taste characteristic of beefsteak and yellow types, weigh up to 2½ pounds. Not sold through listed catalogues; available in nurseries in the South or through seed savers organizations.

Dombito 67 days. A European hybrid used exclusively for commercial greenhouse growing. Taste is inferior to that of the more common greenhouse varieties, Vendor and Tropic. Determinate. Catalogue 32.

Duke (vf₁f₂) 78 days. A commercial variety developed for mechanical harvest. Not to be confused with The Duke, a home garden variety from Park Seeds in Greenwood, S.C. Hybrid, determinate. Catalogue 34.

Duke, The (vf) 75 days. This home garden beefsteak variety produces huge fruits that weigh up to 2 pounds. The vigorous indeterminate vines grow best in a cage. Not to be confused with Duke, a smaller variety used mainly for commercial growing. Standard. Catalogue 26.

Dwarf Champion 72 days. This old variety, though small, stands so upright on its stout stems that it was formerly known as the "tree tomato." (It in no way resembles the tamarillo, page 80, the fruit promoted in the U.S. as a tree tomato.) It is the forerunner of the modern container variety, Patio, and produces a relatively scanty crop of 4-ounce pink, mild-tasting fruits. The compact bush grows no more than 26 inches tall. Standard, determinate. Catalogues 31, 32.

DX 52-12 (vf) 70 days. Developed in Utah as a processing tomato, but now one of the most commonly grown varieties by home gardeners in that state. Large, firm, good-flavored fruits are said to be as good for slicing as for canning. They ripen evenly, maturing at the same time on the large determinate vine. Fruit-setting ability is not impaired by temperature extremes, and heavy foliage prevents sunscald. Less subject to blossom end rot than most varieties. Standard. Catalogue 23.

Early Boy 66 days. This determinate hybrid is best suited to the southeastern states. Thick-walled fruits weigh 8 to 9 ounces and have virtually no cores. Foliage is an unusual dark green. Not sold through listed catalogues; available at nurseries in the Southeast.

Early Detroit 78 days. A pink tomato that, despite its name, matures late. Bushy determinate plants produce large, firm, pink fruits with mild flavor. Standard. Catalogue 32.

Early Pak 7 81 days. The "early" in the name of this variety is misleading, since maturation time is quite long; the "pak" suggests correctly that the variety is intended to be packed and shipped green to market. Nevertheless, the tomato is often grown by home gardeners from California to Colorado for its high yields and the suitability for canning of its thick-walled, 8-ounce fruits. Large determinate, standard. Catalogues 5, 29.

Early Pak 707 (VF)   81 days. An improved version of Early Pak 7, with smoother fruits of more uniform size. Standard. Not sold through listed catalogues; available at nurseries in the West.

Eilon (VF)   78 days. This Israeli import adapts best to the South, but has also been shown to perform well in the Midwest. Large determinate vines produce 4- to 5-ounce fruits of good flavor. Standard. Catalogue 35.

Everbearing, Burgess   74 days. This variety continues to produce tomatoes on the branches long after the crown set has been picked. Fruits weigh up to 11 ounces and have mild flavor. Dense foliage gives good protection from sunscald and insulates fruits from cold snaps. Standard. Catalogue 3.

Fantastic   70 days. Page 58.

Flora-Dade   78 days. Page 59.

Floradel (F)   80 days. A Florida commercial variety that is one of the most common supermarket tomatoes along the eastern seaboard. Like all such varieties, it will have better taste and texture when grown at home and left to ripen fully on the vine. Still, it is inferior to varieties bred specifically for home gardens. Gardeners in humid areas may want to try it because of its resistance to foliage diseases. Six-ounce fruits are firm; indeterminate vines have good foliage cover and are resistant to fusarium wilt and gray leaf spot. Used extensively for commercial greenhouse production. Standard. Catalogue 28.

Floralou   74 days. Similar in quality to Floradel, this hybrid was bred for commercial growers in Louisiana. Fruits are smaller, weighing about 5 ounces, but yield is higher. Resistant to cracking and blossom end rot. Indeterminate. Not sold through listed catalogues; available at nurseries in Louisiana.

Floramerica   72 days. Page 60.

Freedom (VF_1F_2)   75 days. Because this firm commercial variety sets fruit at extremes of temperature, it is a suitable choice for home gardeners in the Southwest and the far northern states and Canada. Jointless 8-ounce fruits are produced in abundance on the large determinate vines. Hybrid. Catalogues 28, 34.

Gardener's Delight   66 days. This cherry tomato is also known as Sugar Lump. Gardeners rate it high for abundant yields and sweet taste. Fruits, measuring 1½ inches and borne in clusters of 6 to 12, ripen evenly to the stem and are resistant to cracking. Standard, determinate. Catalogues 4, 18, 35 (as Gardener's Delight); 19, 26, 29 (as Sugar Lump).

Giant Belgium   90 days. An enormous tomato that, despite its name, was bred in Ohio. Dark pink fruits average 2 pounds, but have been known to grow as large as 5 pounds. Like most beefsteak types, Giant Belgium is low in acid and has a sweet taste. Indeterminate, standard. Catalogue 11.

Giant Climbing, Jung's   85 days. An exclusive of the J. W. Jung Seed Company of Wisconsin, this potato-leaved variety is similar to Climbing Trip-L Crop. Fruits weigh 1 pound or more, are deep red in color, and ripen evenly to the stem. Many, however, are subject to catfacing and may be grotesquely misshapen. Tall indeterminate vines are best grown on a trellis or in a cage. Standard. Catalogue 19.

Giant King   80 days. Introduced in the 1950s, this large-fruited indeterminate is an old-timer by hybrid standards. It has been superseded by higher-yielding, disease-resistant hybrids such as Better Boy. Catalogue 9.

Giant Tree   90 days. This old potato-leaved variety is almost identical to Climbing Trip-L Crop, producing 1- to 2-pound mild fruits on a vigorous in-determinate vine. Not to be confused with the tamarillo (page 80), which is promoted as a tree tomato. Standard. Catalogues 14, 29, 33.

Glamour   78 days. Crack resistance is this variety's main recommendation. Firm, 6-ounce tomatoes are borne on sparsely foliaged vines that are best grown in a cage. Performs well in the northeastern states and eastern Canada. Standard, indeterminate. Catalogues 3, 7, 13, 25, 30, 32.

Godfather (vfn)   75 days. An indeterminate hybrid that dates from 1975. Large fruits weigh about 10 ounces, but often grow up to 1 pound. Plants offer good foliage cover and are resistant to verticillium wilt, fusarium wilt, and nematodes; they also offer some resistance to gray leaf spot, gray wall, and blossom end rot. Catalogue 29.

Golden Boy   78 days. Page 66.

Golden Queen   79 days. Introduced in 1882, this is one of the few yellow heirloom varieties still available through a seed catalogue. It was the most popular large-fruited yellow with home gardeners until the introduction of Jubilee in 1943. Fruits weigh 6 to 8 ounces and are slightly flattened. Their color is bright yellow, unlike the newer yellow-orange types, and their flavor is mild. Standard, indeterminate. Catalogue 32.

Goldie, Park's   66 days. A yellow cherry tomato for hanging baskets, pots, and window boxes. Dwarf plants grow about 14 inches and overhang the sides of containers. The golden color of the 1-inch fruits makes the plant an attractive and unusual ornamental. Not to be confused with the variety of husk tomato (ground cherry) called Goldie (page 64). Catalogue 26.

Greater Baltimore   81 days. This was the most common tomato for commercial canning until the 1930s, when it was replaced by Marglobe and Rutgers. It is popular among home gardeners who like high-acid tomatoes. Fruits weigh about 6 ounces and are quite meaty. Standard, indeterminate. Not sold through listed catalogues; available at nurseries.

Ground Cherry (Husk tomato)   75 days. Page 63.

Gulf State Market   75 days. This old pink variety was one of the South's most commonplace commercial tomatoes in the 1920s. Vigorous indeterminate vines bear 4-ounce fruits that resist cracking. Standard. Not sold through listed catalogues; available at nurseries in Texas and Louisiana.

Gurney Girl (VF₁F₂NT)   75 days. Gurney's Seed & Nursery of South Dakota claims the taste of this hybrid "will restore your faith in modern plant breeding." Gardeners agree that it is one of the best-tasting of the recently introduced varieties. It also has more disease resistance than most tomatoes—to verticillium wilt, both races of fusarium wilt, nematodes, and tobacco mosaic virus. Smooth fruits weigh 6 to 8 ounces. Indeterminate. Catalogue 12.

Harvestvee (VF)   75 days. This variety is usually grown by commercial growers in the North, yet home gardeners frequently use it because it is one of the best varieties for canning whole. Fruits average about 8½ ounces, and are much firmer than home garden types. Determinate, standard. Catalogue 32.

Hayslip (VF₁F₂)   78 days. A Florida supermarket tomato, bred to have higher yields and larger fruits than the widely grown Flora-Dade. Fruits, weighing 5 to 6 ounces, are suitable for home gardeners, because they are not as hard as Flora-Dade and they have a pleasant flavor when ripened fully on the vine. Determinate, standard. Not sold through listed catalogues; available at nurseries.

Heavyweight (VF₁F₂)   80 days. Huge, mild beefsteak fruits weighing up to a pound are produced by this hybrid. Indeterminate. Catalogue 28.

He-Man   73 days. This hybrid has high yields of large fruits that ripen uniformly. Ten-ounce tomatoes are solid and meaty, and are produced all season long. Indeterminate vines have tolerance for verticillium and fusarium wilts, but not full resistance. Catalogues 10, 21, 29.

Heinz varieties   68–77 days. Page 62.

Highlander (v)   68 days. A Colorado-developed tomato used for commercial canning, market sales, and home gardens. Concentrated fruit set of the determinate vine makes the variety a good choice for home canners. Fruits weigh 4 to 5 ounces. Standard. Catalogue 5.

Homestead   82 days. Page 63.

Husk Tomato (Ground Cherry)   75 days. Page 63.

Hybrid 980 (VF)   75 days. This high-yielding, mild flavored hybrid, known mainly in the Northeast, supports itself well enough to be grown unstaked. Fruits weigh about 5 ounces. Determinate. Catalogue 1.

Hy-X 67 days. This hybrid sets fruit well in semiarid areas of the West. Five-ounce fruits are borne on compact determinate plants that require no staking. Catalogue 10.

Independence (VFL F2) 73 days. This firm, thick-walled hybrid is jointless—bred to pick stem-free when mechanically harvested. Taste is slightly better than that of similar commercial varieties. Determinate. Catalogue 34.

Indian River 75 days. A common misconception about this tomato is that it is a hard-to-find heirloom variety; in fact, it was developed by the University of Florida in 1965 as a commercial variety for warm climates. Fruits are medium-sized, weighing about 6 ounces, are firm, and have no blossom end scar. Indeterminate plants provide good foliage cover and resist the wilt and leaf spot diseases common in humid areas. Standard. Catalogue 3.

Ingegnoli 90 days. A giant Italian variety, with tomatoes up to 3 pounds. Fruits are meaty, with few seeds. Said to perform well in Italy, but criticized by North American gardeners for its low yields and late ripening. Standard, indeterminate. Catalogue 11.

J. Moran 90 days. An old standard variety frequently grown by truck gardeners in the western U.S., particularly in southeastern Colorado. Indeterminate vines produce 6- to 8-ounce fruits of good flavor. Largely superseded by newer, disease-resistant varieties. Catalogues 5, 29.

Jackpot (VFN) 72 days. This extra-firm commercial hybrid was bred for shipping, but is rated high by home gardeners for its heavy yields and good disease resistance. Fruits average about 8 ounces. Determinate. Catalogues 18, 20, 22, 28, 32.

Jet Star 72 days. Page 64.

Jubilee 72 days. Page 64.

Jumbo 80 days. A greenhouse tomato used for commercial growing in the northern states and Canada. It is exceptionally high-yielding, bearing clusters of 5- to 7-ounce fruits. Determinate, hybrid. Catalogues 13, 32, 38.

Jumbo Hybrid, Burgess 80 days. Burgess Seeds bred this king-sized tomato to be rounder than most beefsteak types; fruits measure 4½ inches across and 3¼ inches deep. They weigh up to 2 pounds each, and are touted by Burgess as "extra meaty and extra mild." Medium-tall vines should be staked or grown in cages. Catalogue 3.

Jumbo Jim (VF) 84 days. A beefsteak hybrid. Fruits, meaty and mild-flavored, weigh up to 2 pounds. Indeterminate. Catalogues 1, 11.

Liberty (VF₁F₂) 80 days. A commercial hybrid with high yields. Six-ounce fruits are green-shouldered and hard. Determinate plants are highly disease-resistant—to verticillium wilt, both races of fusarium wilt, early blight, and gray leaf spot.

Fruits are not as flavorful as those of home garden varieties. Catalogue 34.

Little King (VF₁F₂N)   65 days. One of the most disease-resistant cherry tomatoes. Indeterminate vines, which should be staked, are resistant to verticillium and fusarium wilts, nematodes, tobacco mosaic virus, and early blight. Fruits hang in clusters, like Sweet 100, but are not as flavorful. Hybrid. Catalogue 26.

Longkeeper   78 days. Page 66.

Lucky Draw   72 days. This hybrid produces 8-ounce crack-resistant fruits on large determinate vines with good foliage cover. Catalogue 28.

Mammoth Wonder   75 days. A beefsteak tomato that is rounder than most of its type, measuring 4¼ inches across and 3¼ inches deep. Firm fruits are crack resistant and weigh up to 1¼ pounds. Standard, indeterminate. Catalogue 3.

Manalucie (F)   85 days. This southern specialty is resistant to sunscald, cracking, and blossom end rot, as well as to fusarium wilt. Meaty, 7-ounce fruits are praised by gardeners for their excellent taste. Standard, indeterminate. Catalogues 15, 26.

Manapal (F)   80 days. A Florida greenhouse variety with hard, 6-ounce fruits. Manapal is sometimes grown by home gardeners because it sets fruit well under adverse conditions. Standard, indeterminate. Catalogues 7, 28.

Marbon   68 days. This tomato is often used for commercial canning and fresh market sales, but is grown by home gardeners, too, because it stands up to changeable weather better than most varieties. Medium-sized fruits have a rich, red color. Determinate, standard. Catalogue 5.

Marglobe   73 days. Page 68.

Marion (F)   70 days. This standard variety is grown mainly in the South. Smooth fruits average 7 to 8 ounces, borne on indeterminate vines with good foliage cover. Catalogue 26.

Marmande   74 days. Page 68.

Marzano Lampadina Extra   80 days. A paste tomato imported from Italy and prized for its fine taste. Determinate vines produce large yields of 2- to 3-ounce fruits that peel easily. Grown for sauce and catsup-making. Standard. Catalogue 8.

Michigan-Ohio (F)   78 days. This hybrid, also known as Michigan-Ohio Forcing, is one of the most commonly grown greenhouse varieties in the Midwest. Vigorous indeterminate vines produce thick-walled 8-ounce fruits. The plants are resistant to fusarium wilt, but many greenhouse growers find them susceptible to leaf mold. Catalogue 3.

Mighty Boy   65 days. A high-yielding pink hybrid that is commonly grown in greenhouses in Canada, but also performs well outdoors in cool climates. Smooth fruits are crack-resistant and average about 7 ounces each. The variety's supplier in Canada says it needs more fertilizer than most other hybrids. Determinate. Catalogue 36.

Minibel   65 days. A dwarf ornamental bred for hanging baskets and window boxes. Bright red, large cherry tomatoes, almost 2 inches across, stand out against cascading dark green foliage. Gardeners report the taste is better than that of most ornamentals. Standard, determinate. Catalogues 26, 29, 37.

Moira   70 days. Developed as a commercial tomato for canning, juice, and fresh-market sales, Moira is nevertheless popular with home gardeners in northern areas as a "bush beefsteak" type—that is, a meatier-than-usual tomato borne on a compact determinate vine. Fruits average 6 to 7 ounces, and are notable for the intense red color of their interiors. Resistant to cracking and blossom end rot. Standard. Catalogues 18, 32.

Moneymaker   70 days. This variety is said to have been so named because English greengrocers could hardly keep up with the demand for its well-formed, flavorful fruits. The plants thrive in cool weather, setting fruit better in spring or fall than in the heat of summer. Four-ounce tomatoes are borne on indeterminate vines. Standard. Catalogue 12, 35.

Monte Carlo (VFN)   75 days. Gardeners call Monte Carlo one of the better-tasting new hybrids. Its 9-ounce fruits are produced all season long on strong indeterminate vines that provide good foliage cover. Catalogues 9, 15, 32.

Moreton   70 days. Page 69.

Napoli (VF)   75 days. Pear-shaped paste tomatoes, very firm and averaging 2½ ounces, are produced abundantly on a compact, determinate plant. Napoli has a more concentrated fruit set than the popular Roma. Standard. Catalogues 7, 11, 28.

Nematex (FN)   75 days. This Texas-bred tomato has resistance to most species of nematodes, and grows well throughout the Southwest, Nevada, and most of California. Seven-ounce fruits are rather hard, but have a pleasant, mellow flavor when picked dead ripe. Determinate vines provide good foliage cover. Standard. Not sold through listed catalogues; available at nurseries in the Southwest.

Nova   65 days. This is the earliest of the paste tomatoes, and is commonly grown in areas with short growing seasons. In other respects, it is similar to Roma, with plum-shaped, 2-ounce fruits used for sauces, catsup, and tomato paste. Determinate vines are compact. Standard. Catalogues 18, 32.

Ohio MR 13 Pink   70 days. A pink greenhouse variety, praised by commercial growers because it ripens more evenly than many forcing types. Fruits average 8 ounces and are generally crack-free. Resistant to tobacco mosaic virus. Standard, determinate. Catalogue 32.

Olympic (F)   76 days. A pink tomato, common in and around Montreal. The firmness of its 8-ounce fruits makes them more suitable for commercial growers than for home gardeners. Resistant to cracking. Indeterminate vines should be staked. Standard. Catalogue 32.

Ontario Hybrid Red 775   74 days. Canadian greenhouse variety. Eight-ounce fruits ripen from the inside, as do those of the better-known Vendor variety, but set at lower temperatures. Determinate vines are immune to all of the races of leaf mold and tobacco mosaic virus that occur in Ontario. Catalogue 32.

Oogata Fukuyu (VFN)   76 days. Excellent taste is the main recommendation of this Japanese import. Meaty 8-ounce fruits are an unusual fluorescent pink when ripe, and have skin that is thinner than that of most varieties. Indeterminate, hybrid. Catalogue 35.

Oxheart   86 days. Page 70.

Pakmore (VF)   75 days. A California-developed tomato popular with small-scale commercial growers in California, Arizona, and Colorado. The variety is similar to the older Pearson tomato, but with higher yields of 7- to 8-ounce fruits. Determinate plants are large enough to need the support of a stake or cage. Standard. Not sold through listed catalogues; available at nurseries in California, Arizona, and Colorado.

Park's Greenhouse Hybrid 130 (VFNT)   70 days. An indeterminate greenhouse variety developed for winter production in the South. Fruits are large, weighing up to 9 ounces. Catalogue 26.

Patio   70 days. Page 70.

Patio Prize   67 days. Page 70.

Pearson Improved (VF)   85 days. There are at least fifteen variations on the original Pearson, a standard developed in California in 1936. The most widely available nowadays is Pearson Improved, which is grown as a fresh-market tomato in the West. Tough-skinned fruits weigh about 8 ounces. Determinate. Catalogues 5, 28, 29.

Pelican (FN)   70 days. A Louisiana specialty, bred to set fruit in high temperatures, to resist the fusarium wilt common in humid areas, and to withstand attack by nematodes. Smooth fruits weigh about 8 ounces. Indeterminate vines should be caged or staked. Standard. Not sold through listed catalogues; available at nurseries in Louisiana.

Perfect Peel (VF)   80 days. A bush-type variety with 6-ounce fruits that peel easily. Ideal for home canners. Hybrid. Catalogues 29, 37.

Peron   68 days. A South American import (not to be confused with Perron 50, a Canadian variety sold by the W. H. Perron Company) claimed to be so pest-resistant that it never needs spraying. Eight-ounce fruits have good flavor and are higher in vitamins C and A than most other varieties. Indeterminate. Not sold through listed catalogues; sometimes available at nurseries or through seed savers organizations.

Pik-Red (vF₁F₂)   71 days. A hybrid notable for its almost solid interiors. This characteristic makes it desirable for commercial growers, since fruits are so firm they can be picked and shipped red instead of green. Home gardeners may find them too hard. Dwarf determinate plants can be grown close together— an advantage for gardeners with little space. Fruits weigh 6 to 7 ounces. Catalogue 13.

Pink Delight (F)   70 days. This hybrid, developed at the University of Missouri, is rated excellent all around by gardeners. Seven-ounce pink fruits are crack-resistant and exceptionally flavorful. Indeterminate. Catalogues 2, 3, 12, 34, 37.

Pink Gourmet (F)   72 days. This pink beefsteak was bred to mature earlier than most beefsteak types. Fruits, borne on vigorous indeterminate vines that should be staked or caged, weigh up to 1 pound or more. They are meaty and mild, with the good flavor characteristic of older pinks such as Ponderosa. Hybrid. Not sold through listed catalogues; available at nurseries in Missouri and surrounding states.

Pink Savor   75 days. This small plum-shaped tomato, developed in Missouri for use in salads and home canning, is slightly larger than most cherry tomatoes and is rated high by gardeners for its mellow flavor. Indeterminate vines should be staked, but not pruned. Standard. Not sold through listed catalogues; available at nurseries in Missouri and surrounding states.

Pinkshipper (F)   74 days. A pink commercial variety widely grown in the South in the 1960s. Its large cores and less-than-abundant yields led to its replacement by the superior Bradley variety in commercial fields and home gardens. Fruits weigh about 4 ounces. Determinate, standard. Not sold through listed catalogues; sometimes available at nurseries in the South.

Pole Boy #83 (F₁F₂)   80 days. Indeterminate commercial variety. Eight-ounce fruits hold their size throughout the growing season, but are inferior to home garden types in taste and texture. Standard. Catalogue 34.

Pole King (vF₁F₂)   75 days. This commercial variety has exceptionally high yields, large fruits, and good disease resistance. Fruits, weighing up to 10 ounces, are thick-walled and have good keeping quality. Indeterminate vines are able to set fruit in unseasonably cold or hot weather. Hybrid. Catalogues 34, 38.

Ponderosa   90 days. Page 71.

Porter   70 days. Page 72.

Porter's Pride   70 days. Page 73.

President (VFNT) 68 days. This hybrid is one of the few determinates developed especially for the home garden. Although the fruits, weighing 6 to 8 ounces, are meaty and have good flavor, disease resistance is the variety's main recommendation. Large determinate plants should be supported on short stakes or in cages. Catalogues 9, 21, 29.

Pritchard (F) 70 days. Also known as Pritchard's Scarlet Topper, this old variety was developed by the United States Department of Agriculture in 1931. Mild fruits average about 5 ounces and are protected by a heavy cover of leaves. Standard, determinate. Catalogues 14, 31.

Quick Pick (VFNT) 66 days. Disease resistance is the best feature of this variety. Gardeners describe its yield and taste as only fair. Five-ounce, meaty fruits are borne on indeterminate vines. Hybrid. Catalogues 15, 26, 29.

Ramapo (VF) 80 days. An indeterminate hybrid bred to set fruit under adverse conditions. Thick-walled fruits also resist cracking and blossom end rot. Catalogues 7, 13, 22.

Red Cherry 72 days. Page 73.

Red Chief (VFN) 80 days. This southeastern specialty, from the H. G. Hastings Company of Atlanta, is an indeterminate, home garden hybrid bred to set fruit in hot weather. Smooth tomatoes weigh about 9 ounces and measure 3 inches across. The vigorous vines should be staked or caged. Catalogue 15.

Red Express 238 (VFN) 74 days. A determinate hybrid producing large fruits weighing 8 to 10 ounces. Although bred primarily for firmness and long shelf life, the tomatoes have flavor comparable to home garden types. Catalogue 18.

Red Glow (VFN) 72 days. This medium-sized tomato is more outstanding for its market attributes—perfect round shape and intense red color—than for its taste. Indeterminate vines produce large yields of 6-ounce fruits. Grows best east of the Rocky Mountains. Catalogue 34.

Red King (FT) 73 days. Catalogues describe Red King as "perhaps the most solid tomato ever developed," a clue to home gardeners that it is rock-hard. It is bred to stay firm after being picked ripe from the vine—a quality necessary for shipping and for display in roadside stands, but unnecessary for home gardeners who prefer succulence over keeping quality. Fruits weigh about 6 ounces. Standard, determinate. Catalogue 34.

Red Mountain 75 days. This hybrid, developed especially for cool mountain areas in the West, produces mild-tasting, 8-ounce fruits. Indeterminate. Catalogue 29.

Red Peach 66 days. Small-fruited variety used as an ornamental or for preserving and pickling. Fruits weigh about 2 ounces and measure 2 inches across. Determinate. Catalogue 32.

Red Pear   70 days. Page 84.

Red Plum   70 days. Page 85.

Red Queen (VF₁F₂N)   78 days. A western commercial hybrid with firm, 6½-ounce fruits. Indeterminate vines have good foliage cover to protect fruits from sunscald. Catalogue 7.

Redheart (F)   75 days. This Missouri-bred hybrid is similar to Big Boy, with fruits weighing up to 1 pound, and is named for the meaty, deep red flesh at the tomatoes' centers. Indeterminate vines produce abundantly throughout the growing season. Catalogue 21.

Roadside Red (VF)   73 days. The 5- to 6-ounce fruits of this variety ripen evenly and are free of cracks and blemishes, making them ideal for display at roadside stands or in markets. They are satisfactory for eating fresh and for home canning. Hybrid, determinate. Catalogue 1.

Rockingham   70 days. This New Hampshire-bred tomato was developed to resist late blight. Fruits weigh 6 to 8 ounces. Indeterminate, standard. Not sold through listed catalogues; available at nurseries in northern states.

Roma   75 days. Page 74.

Roseplus   68 days. A pink Canadian hybrid that ripens without green shoulders. Large determinate vines produce 5-ounce fruits that resist cracking. Usually grown commercially. Indeterminate. Catalogue 38.

Royal Flush (VFN)   68 days. A West Coast commercial hybrid. Determinate plants yield 8-ounce fruits, measuring up to 3 inches across, in a concentrated set. Catalogue 20.

Rushmore (VF)   67 days. Developed to set fruit in the cool springs and hot summers of the Dakotas, Rushmore has high yields of meaty, firm 7-ounce tomatoes. Gardeners rate them especially high for taste. Hybrid, determinate. Catalogues 12, 32.

Rutgers   80 days. Page 75.

Saladette (F)   65 days. This Texas-bred variety sets fruit in high temperatures and produces a heavy yield of small, plum-shaped salad tomatoes that average about 2 ounces. Fruits are hard, thick-walled, and crack-resistant. Compact determinate vines spread only 1½ feet. Standard. Not sold through listed catalogues; available at nurseries in the Southwest, California, and Nevada.

San Marzano   80 days. Page 75.

Show Me (F)   70 days. This Missouri-bred hybrid is grown both as a commercial and home garden tomato. Firm 8-ounce fruits keep well for several days on the vine after they are fully ripe. Determinate. Catalogue 21.

Small Fry   65 days. Page 76.

Spring Giant   66 days. Page 77.

Springset   65 days. Page 77.

Square Paste   74 days. So-called square tomatoes are used for processing. Their shape keeps them from cracking when they are transported in bulk from machine harvester to processing factory. They are of interest to home gardeners only as curiosities, since the tomato paste and sauce they make is less flavorful than that of older paste varieties. Standard, determinate. Catalogues 11, 32.

Stakebreaker (vf)   75 days. Named for its high yields, this vigorous hybrid grows well along the East Coast from Virginia to Florida. Large fruits, averaging about 10 ounces, do not decrease in size at the end of the season. Indeterminate. Catalogue 15.

Stakeless (f)   78 days. This variety is a tree type, with an 18-inch vine that supports itself on stout stems. Fruits weigh 6 to 8 ounces and are quite mild. The plant is suitable for growing in containers, but needs to be staked when fruits begin to mature. In the garden, it needs no staking. Standard, determinate. Catalogues 5, 10, 12, 32, 34.

Star-Pak (vf_1f_2)   79 days. This high-yielding indeterminate hybrid was developed in New York State, and performs well east of the Mississippi. Fruits weigh up to 9 ounces, measure 3 inches across, and are fairly firm. Catalogue 13.

Stokes Pak (vfn)   67 days. An extra-firm commercial variety common in the northern states and Canada. Hybrid, determinate. Catalogue 32.

Stokesdale   72 days. High-yielding determinate variety for northern areas. Fruits weigh up to 8 ounces and have rich red interiors. Standard. Catalogue 32.

Stone   84 days. This old tomato dates from 1889 and has largely been superseded by newer varieties. Its 6- to 7-ounce fruits are used mostly for canning. Standard, determinate. Catalogue 3.

Stuffing Tomato, Burgess   78 days. This hollow variety, shaped more like a bell pepper than a tomato, is ready-made for stuffing with tuna salad or other fillings. Fruits measure 3¼ inches across and 2¾ inches deep. Standard, determinate. Catalogues 3, 11, 16.

Summertime, Improved (fn)   75 days. A Texas home garden variety, bred to set fruits during the hot months of July and August. Fruits are small, measuring up to 2½ inches across, and are crack-free. Compact determinate vines offer good foliage cover to prevent sunscald. Standard. Catalogue 28.

Sunlight (vfn)   70 days. This large indeterminate is notable for its resistance to foliage diseases. Fruits weigh 6 to 8 ounces. Hybrid. Catalogue 28.

108

Sunray   78 days. Page 66.

Sunripe (VFN)   75 days. This large-fruited determinate performs best in the western half of the continent. Compact plants, grown on short stakes or in cages, produce a good yield of 8- to 10-ounce fruits. Hybrid. Not sold through listed catalogues; available at nurseries in the West.

Sunset   65 days. A northern commercial variety similar to Starfire and Fireball. Although determinate, it continues to bear fruit through the middle of the growing season, and, in some cases, until frost. Fruits weigh about 4 ounces. Standard. Not sold in listed catalogues; available at nurseries in New England.

Super Bush (VFN)   70 days. This home garden variety was bred to require little care on the gardener's part: it needs no staking, pruning, or caging, has good disease resistance, and produces 5-ounce fruits throughout the growing season. Plants grow about 3 feet tall and 3 feet wide, making them a good choice for urban gardeners who must grow tomatoes in containers. Hybrid, determinate. Catalogue 30.

Super Fantastic   70 days. Page 58.

Super Red (VF)   85 days. An indeterminate hybrid for home gardens in the Northeast. Fruits are large, averaging 8 to 10 ounces, and are an intense red. Catalogue 1.

Super Sioux   70 days. Page 78.

Supermarket   75 days. A Texas variety that was bred for grocers' shelves rather than home gardens. Fruits weigh 5 to 6 ounces and are uniform in color. Determinate, standard. Not sold through listed catalogues; available at nurseries in Texas.

Supersonic (VF)   79 days. A northeastern variety that produces large crops of crack-resistant, meaty 9-ounce tomatoes. Vigorous determinate vines are best grown in a cage. Hybrid. Catalogues 2, 13, 14.

Superstar   80 days. This West Coast specialty is rated high for its meatiness and good flavor. Fruits weigh up to 10 ounces and measure 3½ inches across. Hybrid, indeterminate. Catalogue 17.

Supersteak   80 days. Page 79.

Supreme (F)   70 days. The actual name of this Missouri-bred variety is Mocross Supreme, but the first word is frequently omitted in seed catalogues and at nurseries. The 5- to 7-ounce fruits are rather hard and have green shoulders, but are produced abundantly for most of the season on the large determinate plants. Hybrid. Not sold through listed catalogues; available at nurseries in Missouri and surrounding states.

Surprise (F)   68 days. Also known as Mocross Surprise, this indeterminate hybrid produces an enormous yield of 8-ounce fruits. Unlike the similar Supreme, the fruits do not have green shoulders. Although the tomato was developed as a commercial variety, it is popular among Midwestern home gardeners, who say it develops excellent flavor if left to ripen fully on the vine. Catalogues 2, 7, 10, 21.

Sweetie   62 days. Sweetie is rather like an open-pollinated version of the hybrid Sweet 100, with clusters of exceptionally sweet cherry tomatoes borne profusely in clusters. Yields are not as high, however. The bite-sized fruits have more vitamin C than most tomatoes. Standard, indeterminate. Catalogues 10, 15, 24, 39.

Sweet 100   66 days. Page 79.

Tamiami (vF₁F₂)   82 days. A Walter-type commercial variety that is widely grown in humid areas because of its resistance to leaf diseases. Determinate vines with good foliage cover produce hard 7- to 8-ounce fruits. Resistant to gray leaf spot as well as verticillium wilt and two races of fusarium wilt. Standard. Catalogue 34.

Terrific (vFN)   73 days. This home garden hybrid is noted for its excellent taste. Strong indeterminate vines produce smooth 10-ounce fruits over a long season. Good foliage cover and ability to set fruit in hot weather make Terrific a popular choice for southern gardeners. Catalogues 6, 15, 26, 28.

Tomboy   70 days. This midwestern variety is called a beefsteak type because of the meatiness and roughness of its fruits; but while they are large, averaging 10 to 12 ounces, they are smaller than most beefsteaks, and ripen much earlier. Standard, indeterminate. Catalogues 2, 21.

Toy Boy (vF)   68 days. An ornamental cherry variety with 1-inch fruits on a plant that grows no more than 12 to 14 inches tall. It is attractive in hanging baskets or pots, but gardeners say its taste is blander than that of most cherry types. Hybrid, determinate. Catalogues 1, 6, 14, 24, 25, 26, 28, 32, 38.

Tropic (vFN)   80 days. A popular greenhouse variety with excellent disease resistance—to early blight and gray leaf spot as well as verticillium wilt, fusarium wilt, and nematodes. Firm fruits weigh 8 to 9 ounces. Standard, indeterminate. Catalogues 4, 27, 28, 29, 32.

Ultra Boy (vFN)   72 days. This indeterminate hybrid is more suited to commercial growing than the other "Boy" tomatoes, since it has higher yields and better keeping quality, but is less satisfactory for home gardens because of its hardness and inferior taste. Fruits weigh 1 pound or more. Catalogue 32.

Valiant   66 days. An old commercial variety similar to Earliana. Six-ounce, mild-flavored fruits ripen evenly to a rich red. Sold as Field's Red Bird in the catalogue of Henry Field Seed & Nursery of Shenandoah, Iowa. Standard, indeterminate. Catalogues 3, 6, 7, 10, 28, 33.

Veepick (vf)   73 days. An elongated paste tomato that measures 3¼ inches long and 2 inches thick. It peels easily for use in sauce, catsup, or tomato paste. Standard, determinate. Catalogue 32.

Veepro (v)   70 days. A paste variety bred for mechanical harvest. Hard fruits are crack-resistant. Standard, determinate. Catalogue 36.

Veeroma (vf)   72 days. A paste tomato used only for catsup and tomato paste. It is adapted to mechanical harvest and is higher yielding than the popular Roma. More suitable for commercial growers than for home gardeners. Standard, determinate. Catalogues 32, 38.

Vendor   66 days. A greenhouse variety with 6- to 8-ounce fruits. Largely superseded by Tropic, which has better disease resistance. Standard, indeterminate. Catalogues 18, 32, 38.

Walter   75 days. Page 81.

Westover   75 days. A determinate commercial variety with good foliage cover and a concentrated fruit set. Fruits weigh 5 to 7 ounces and ripen uniformly. Standard. Catalogue 22.

White Beauty   84 days. Page 82.

White Wonder   85 days. Page 82.

Whopper, Park's   75 days. Page 82.

Willamette   67 days. An Oregon specialty, noted for its earliness, crack resistance, and good flavor. Tomatoes harvested early in the season average about 5 ounces, but decrease in size as the season progresses. Standard, determinate. Catalogue 24.

Wisconsin 55   75 days. Developed by the University of Wisconsin, this commercial variety produces fleshy, 6-ounce tomatoes that ripen evenly and are of uniform size. Determinate vines have tolerance for several leaf diseases, including early blight and gray leaf spot. Standard, determinate. Catalogues 19, 25.

Wonder Boy   80 days. Page 83.

Yellow Cherry   72 days. Page 74.

Yellow Peach   67 days. A small yellow tomato, weighing about 2 ounces and measuring 2 inches across, that is used exclusively for pickling and preserving. Compact determinate vines are suitable for growing in containers, and make a fine ornamental display when grown with Red Peach. Standard. Catalogue 32.

Yellow Pear   76 days. Page 84.

Yellow Plum   70 days. Page 84.

TOMATOES FOR SPECIAL PURPOSES

The following lists are provided as quick references for gardeners who are looking for varieties of certain size, color, use, or weather tolerance.

LARGE-FRUITED VARIETIES

These are king-sized tomatoes, sometimes called "sandwich tomatoes" because one slice almost covers a piece of bread. All of the varieties listed here yield fruits that weigh at least 12 ounces; many have 1-pound fruits, and some will grow to 2 pounds or more with plenty of water and fertilizer.

Abraham Lincoln, page 38

Beefmaster, page 42

Beefsteak, page 41

Better Boy, page 43

Big Boy, page 44

Big Girl, page 94

Big Pick, page 45

Big Sandy, page 94

Big Seven, page 94

Bragger, page 49

Brandywine, page 49

Celebrity, page 53

Climbing Trip-L Crop, page 55

Colossal, page 96

Delicious, page 56

Dixie Golden Giant, page 97

Giant Belgium, page 98

Giant Climbing, Jung's, page 99

Giant Tree, page 99

Godfather, page 99

Heavyweight, page 100

He-Man, page 100

Ingegnoli, page 101

Jubilee, page 64

Jumbo Hybrid, Burgess, page 101

Jumbo Jim, page 101

Mammoth Wonder, page 102

Oxheart, page 70

Pink Gourmet, page 105

Ponderosa, page 71

Ramapo, page 106

Redheart, page 107

Super Beefsteak, page 42

Super Fantastic, page 58

Supersonic, page 109

Terrific, page 110

Ultra Boy, page 110

Whopper, Park's, page 82

Wonder Boy, page 83

SMALL-FRUITED AND CHERRY VARIETIES

These tomatoes, ranging from 1 inch to 2½ inches in diameter, are generally used in salads. Small-fruited paste types are listed separately.

PASTE VARIETIES

Paste types, usually small and plum-shaped, are generally used for processing—for making sauce, catsup, or tomato paste, or for canning whole. They have meaty, solid flesh, but are rarely eaten fresh because they lack the succulence of other types.

HOME CANNING VARIETIES

Any tomato can be canned, but some are better for the purpose than others. Desirable qualities include meatiness, even ripening, and small cores (or no cores at all). The varieties listed here have those characteristics. Most are determinate: their concentrated production means the home canner will reap a large harvest of fruits over just a few days. Many of the previously listed plum and paste types are also good for canning.

Big Early, Burpee's, page 85

Burpee's VF, page 95

Campbell 1327, page 51

Crack-Proof, page 96

Doublerich, page 53

DX 52-12, page 97

Earlibright, page 86

Early Boy, page 97

Early Pak 7, page 97

Early Pak 707, page 98

Early Pick, Burpee's, page 87

Easy Peel, page 88

Greater Baltimore, page 99

Harvestvee, page 100

Heinz 1350, page 62

Highlander, page 100

Marbon, page 102

Moira, page 103

Pink Savor, page 105

Plainsman, page 90

Red Pear, page 84

Red Plum, page 85

Roadside Red, page 107

Roza, page 91

Rutgers, page 75

Spring Giant, page 77

Springset, page 77

Sprinter, page 91

Stokesdale, page 108

Sun-Up, page 92

Sunripe, page 109

Super Sioux, page 78

Vigor Boy, page 92

Wayahead, Improved, page 92

Westover, page 111

Yellow Pear, page 84

Yellow Plum, page 84

CONTAINER VARIETIES

While all tomatoes, even the indeterminate varieties, will grow in containers, some are bred especially for small plant size and tidiness. These are the determinate bush types, and include miniature and dwarf varieties that are usually grown as ornamentals in pots, window boxes, or hanging baskets. A few of the larger determinates may require the support of a short stake or trellis.

Basket King, page 40

Basket Pak, page 93

Bitsy, page 85

Bonner, page 86

Cherry Grande, page 86

City Best, page 86

Cocktail, page 96

Dwarf Champion, page 97

Early Cherry, page 87

Early Salad, Burgess, page 87

Early Temptation, page 88

Florida Basket, page 88

Florida Petite, page 61

Gem State, page 88

Goldie, Park's, page 99

VARIETIES FOR HOT CLIMATES

Most varieties refuse to set fruit when day temperatures are above 90° and night temperatures above 75°. The following were bred to stand up to summer heat, and are grown in the South and Southwest.

VARIETIES WITH GOOD FOLIAGE PROTECTION

Dense foliage insulates tomatoes from the sudden extremes of temperature that can affect fruit-setting ability, and protects the fruits from sunscald. These varieties provide a good cover of leaves.

VARIETIES FOR HUMID AREAS

High humidity and higher-than-average rainfall make most varieties susceptible to fungus diseases that can defoliate tomato plants. The following varieties have inbred resistance to the most common of those diseases.

Atkinson, page 40

Big Pick, page 45

Calypso, page 95

Campbell 1327, page 51

Caribe, page 95

Contessa, page 96

Creole, page 96

Floradel, page 98

Floralou, page 98

Indian River, page 101

Liberty, page 101

Pelican, page 104

GREENHOUSE VARIETIES

Tomatoes grown in greenhouses receive less sunlight than those grown outside. The following varieties were bred to grow well under such conditions. Many of them will also thrive outdoors in the garden. Most are commercial varieties grown for market sales in winter, but will be of interest to greenhouse hobbyists.

Burpeeana Early, page 86

Burpee's VF, page 95

Crimson Trellis, page 96

Dombito, page 97

Floradel, page 98

Jumbo, page 101

Manapal, page 102

Michigan-Ohio, page 102

Mighty Boy, page 102

Ohio MR 13 Pink, page 103

Ontario Hybrid Red 775, page 104

Park's Greenhouse Hybrid, page 104

Tropic, page 110

Vendor, page 111

PINK VARIETIES

Pink tomatoes are generally meaty and mild-flavored—not because of their pigment (most are actually red inside, but look pink because they have translucent skins), but because their high sugar content masks their acidity.

Bradley, page 48

Early Detroit, page 97

Giant Belgium, page 98

Gulf State Market, page 100

MacPink, page 89

Mighty Boy, page 102

Ohio MR 13 Pink, page 103

Olympic, page 104

Oogata Fukuyu, page 104

Oxheart, page 70

YELLOW AND ORANGE VARIETIES

These tomatoes, some of which are the oldest varieties available, range in color from tangerine to bright yellow. Like the pinks, they are meaty and mild-tasting. Their acid content, however, is no lower than that of red tomatoes.

PART THREE
CULTIVATION

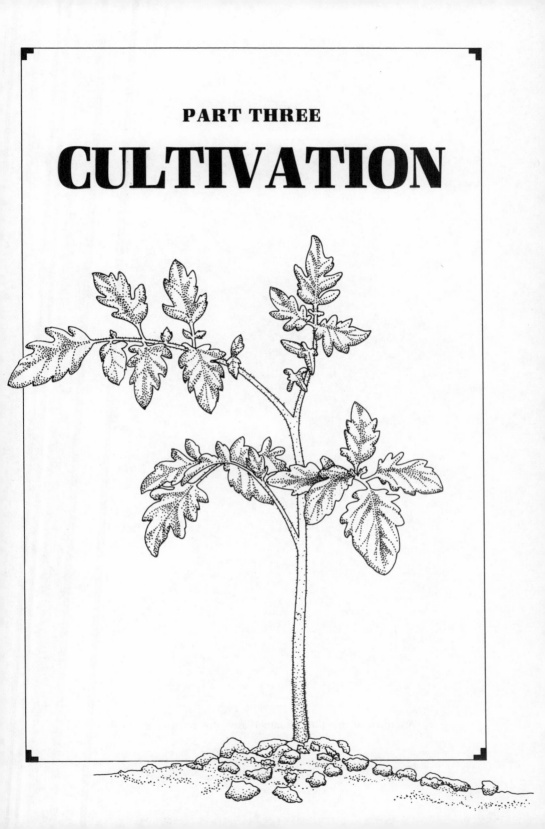

One of the pleasurable things about growing your own tomatoes is that there is no single, fixed way to do it. The rules are not carved in stone. There is a large margin for error, an advantage desirable in any leisurely pursuit. That margin leaves plenty of room for experimenting with cultivation techniques that range from the ordinary to the unorthodox.

People who grow tomatoes year after year tend to put a personal stamp on their gardens, doing things in quite their own way. After a few years of trial and error, they are satisfied that their methods, no matter how unusual, are the surest way to success. And they are eager to share their ideas with other gardeners.

The cultivation instructions on the following pages include both the traditional and the unusual. Standard, time-honored procedures for growing tomatoes are described, and are followed in some cases by the personal, home-developed techniques of individual backyard gardeners. (Such techniques appear in italics). These gardeners are not professional horticulturists, nor scientists, but true grass-roots experts who grow tomatoes every summer and have found approaches that work for them. Not all of the techniques are original. Some are regional folk methods that have been passed down from one generation to the next; others are ingenious modifications of traditional, widely practiced growing methods. All should be of interest to home gardeners, and all are worth a try.

CONSIDERING CLIMATE

Tomatoes are perennials. Yet in North America we treat them as annuals—plants that do not last through the winter and must be planted again at the beginning of each growing season. We rush the plants through their seed-to-seed cycle between the last frost of spring and the first one of fall. And the farther north we live—the farther from the plant's native equatorial region of South America—the harder tomatoes are to grow.

At its most fundamental, tomato growing amounts to outwitting the weather. Twenty years ago, this almost always involved the fairly laborious steps of growing seedlings indoors in winter and protecting them from snap cold spells once they were set out in the garden; in humid areas, it meant spraying plants with fungicides to prevent foliage diseases. Such steps still are necessary for most of the older varieties of tomatoes, and for many of the newer ones. But today, gardeners have an advantage: they can take care of some climate problems simply by growing varieties that are tailor-made for certain temperature ranges and humidities.

The most obvious step is to consider a variety's "days to maturity"—the time it takes to produce ripe fruits. The farther north a gardener lives, the earlier the varieties he should seek. Early types—those that ripen in fewer than 65 days under normal conditions—are the best choices for areas with exceptionally short growing seasons, chilly summer nights, or a lot of overcast days.

A second, more important, consideration is to keep blossoms from dropping off the plant when the temperature rises too high or drops too low. Healthy tomato plants themselves will thrive in summer, quickly outgrowing the stakes or filling the cages that support them, as long as night temperatures do not dip too close to the freezing mark. But the fact that too-cold or too-hot weather will not necessarily stunt tomato plants is cold comfort, since we care less about the size and lushness of the plant than we do about the number of tomatoes on it. When summer night temperatures drop below 55° for a few nights in a row, as they usually do in the northern states and Canada, or rise above 75°, as they often do on sultry nights in the South, the tomato plant's fragile yellow flowers break off and drop to the ground. And each fallen flower means one less tomato; the flower has not had a chance to set fruit.

Plant breeders have worked to give us temperature-adaptable varieties of tomatoes—varieties that will keep their blossoms at extremes of temperature. Strangely enough, many of the varieties that set fruit at very low temperatures will also set fruit at very high ones. Gardeners who live in hot or cold climates can greatly increase their yields, and lessen their worry, by choosing those varieties that have been bred to hold on to their blossoms under adverse conditions. (For other factors that can cause blossom drop, see page 169.)

Another climate concern is humidity. Constant wet weather or a run of muggy days can leave tomato plants susceptible to fungus diseases such as early blight, anthracnose and gray leaf spot. But plant breeders have again come to the rescue with a number of varieties bred to resist the fungus damage encouraged by high humidity. Science has, in fact, adapted the tomato to so many climates that we can easily find ones that will stand cool mountain summers, blazing southern sun, or even the hot days and chilly nights of the desert.

WHERE TO PLANT?

Choosing a place to grow tomatoes should be easy in all but the most densely shaded yards. First, look for a spot that gets at least six hours of direct sunlight every day. This is simple to determine—all you have to do is look out the window a few times each day—but is also the one factor over which gardeners have little control, since the cutting down of trees is not to be advised, even for the sake of home-grown tomatoes. You probably will be choosing a spot in late winter. January or February is the usual time to plan a garden; it is also the time of year you are most likely to have visions of vine-ripe tomatoes from your own tomato patch, since the quality of supermarket tomatoes is at its absolute nadir. If looking for a spot in late winter or early spring, take into account that the sun will be higher in summer and that the shadows cast by the house, sheds and trees will not be as long.

If no spot in the yard gets the needed six hours, do not give up. Tomatoes will grow in less sunlight, but will take longer to mature. Midseason varieties—those that normally take about 65 to 85 days to ripen—may need 95 days or even more to ripen in shade, and frost may kill them off before then in all but the warmest climates. Early varieties, however—those that usually take only 55 to 65 days to mature—will ripen before the first frost, and the gardener will have fresh tomatoes despite his shady yard. (Early varieties often lack the special succulence of later varieties, but still will be considerably better than the commercial alternative.)

Places in the garden that receive *more* than six hours of direct sunlight may present a problem, too, especially in hot southern climates where prolonged exposure can "sunscald" tomatoes. Gardeners should plant varieties that are leafy enough to shade their own fruits (page 115), or grow their plants in a wire cage or tower that will crowd the leaves and provide a sunscreen.

Once you have established which parts of the yard are sunny enough for tomatoes, consider how well those places drain. Boggy places, with standing water after a rain, are unsuitable, since tomatoes will curl up and suffocate if their roots are always wet. But if you find that the only suitably sunny spot is one that doesn't drain well, you can solve the problem by mounding up the topsoil into a raised planting bed, which will drain freely (page 140). Or, easiest of all, you can grow certain varieties in containers (page 114) and forget about drainage altogether.

A third consideration is aesthetic. If placing your tomatoes in a sunny, well-drained spot means they will be the focal point of your yard, you may want to reconsider. Although tomatoes were prized as curiosities and grown only as ornamentals in the seventeenth and eighteenth centuries, and although some people might think tomatoes in the front yard bring a refreshing touch of the country to a characterless subdivision, neighbors might object to tomatoes growing curbside among the neatly manicured lawns of a suburban street.

Growing a couple of tomato plants in a flower bed at the side or back of the house can simplify matters. For one thing, the ground is already cultivated; you may have to dig a little deeper to prepare it for tomatoes, but there is no digging through grass. For another, if the flower bed is against a wall of the house that faces south or west, it will form an ideal microclimate for tomatoes, with the wall absorbing the heat of the sun during the day and radiating it out at night. This can be especially helpful in northern areas, where the growing season is short.

Gardeners who have black walnut trees in their yards (black walnuts are common shade trees east of the Mississippi) should be aware that the tree's roots leach a substance into the soil that tomatoes find intolerable. It will cause them to wilt and die. The plants should not be set out within 30 feet of a large black walnut tree, nor within 10 feet of a small one.

SIZING THE PLOT

Once you have chosen a spot, decide how many plants you want to grow. This determines, of course, the size and shape of the tomato patch.

Two plants per adult in the household, and one per child, usually produce a sufficient crop, barring such unforeseen calamities as an insect plague or a brutal heat wave. Following this rule of thumb, the six plants grown for a family of four will provide enough tomatoes for eating out of hand and putting into sandwiches and spaghetti sauce all summer long. Eight to twelve plants for the same family will yield enough extras for canning or making relish, and twelve to sixteen plants will almost feed the neighborhood.

Naturally, there is no horticultural law that says a gardener must grow two plants per person. A single plant in a pot may satisfy one entire family's occasional craving for the taste of a real tomato. And, in some exceptional cases, one vigorous indeterminate plant, with the perfect combination of temperature, sunlight, and rainfall, will grow at a ferocious rate and become a virtual tomato factory, producing more fruits than even a family of four can use. Still, gardeners who are determined to ban supermarket tomatoes from the house for the summer should follow the two-plants-per-adult rule, for safety's sake. And if the yard is shady, they should put in an extra plant or two to compensate for the reduced yield that is sure to be a consequence of the lower light level.

A 12-×-3-foot row will accommodate six staked tomato plants, and allows them to be grown 2 feet apart. (Tomatoes can be grown closer together, but in typical garden soil their yields will not be as high; they will also be harder to prune and fertilize.) Unstaked plants, which will sprawl, should be spaced 3 feet apart, and, of course, require a longer row. Or, unstaked plants can be grown in a rectangular plot, the boundaries of which they are not likely to trespass. As fruits begin to ripen, the whole plot can be mulched with hay or straw to keep the fruits from resting on the soil and rotting.

The richer and deeper the soil, the more the tomato plants that can be squeezed into the row or plot. In the typical garden, with soil dug to a depth of 1 foot or less, and with a perfunctory addition of fertilizer and some organic matter, a tomato plant needs its own 4 square feet of soil from which to draw nutrients and moisture. In gardens that have been dug to 2 or even 3 feet, and enriched with more than the usual amount of organic matter, tomato plants will not have to compete so keenly for what they need, and staked plants can be placed only 1½ feet apart.

The growth habit of the plants is also a consideration when sizing the plot. Dwarf varieties can be placed 1½ feet apart under any circumstances, and are good choices for gardeners with little space. Determinate varieties that are to be staked can also be placed 1½ feet apart. Unstaked determinates need at least 3 feet between them, and unstaked indeterminates may sprawl so

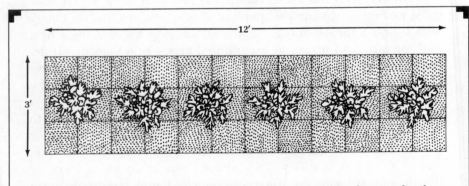

Six staked indeterminate plants will grow best in a row at least 12 feet long and 3 feet wide. A plot this size allows them to be spaced 2 feet apart.

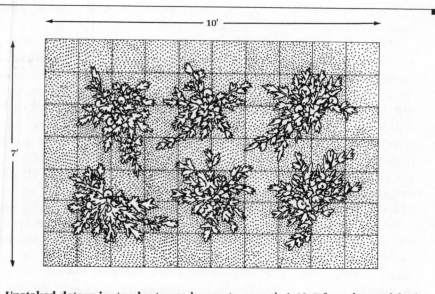

Unstaked determinate plants need room to sprawl. A 10x7-foot plot, mulched to keep the fruits from resting on soil, is adequate for six of the larger determinates such as Marglobe or Heinz 1350. By the time the plants have produced their last fruits, they will have filled the entire plot.

much that they need 4 or even 5 feet between plants, although this makes for grossly inefficient use of space in a home garden. Indeterminates are better staked or grown in a cage.

Gardeners who plan to grow only three or four plants may find it easier to prepare individual planting holes (page 142) than to dig a whole row or plot.

SOIL

Making an effort to identify the type of soil in your garden, and to adjust it to the tomato's taste, will ensure that your plants are more vigorous through the growing season. It will also amount to a minor lesson in geology that can provide an interesting diversion for an afternoon. Simply dig a hole 3 to 4 feet deep, and all you need know about your soil will be revealed: what it is made of, how it holds together, how far down the topsoil goes, and how well it drains after a rain.

Make the hole about 2 feet wide and at least 3 feet deep, trying all the while to keep the walls vertical: the hole should be as wide at the bottom as it is at the top. (It is easier to dig two days after a rain, when the soil is still moist and workable, yet not so muddy that it sticks to the shovel. And keep in mind that digging this hole will probably be the most strenuous work you have to do if you are planting only a few tomatoes rather than digging a full-sized garden for several vegetables. In typical garden soil, planting holes need be only about 1 to 2 feet deep; this deeper test hole may strain the back, but it will pay off in less work for the rest of the season.)

Smooth the walls of the hole with your hands so that you can see the layers of different colors and textures that geologists call the soil profile. There may be only two or three layers, or as many as five or six. Lucky gardeners will find that the top layer, or topsoil, is deep, crumbly, dark and teeming with earthworms. The lower layers, or subsoil, probably will be heavier, with more clay. Measuring all the layers that are loose enough for a plant's roots to penetrate shows the depth of the soil. The deeper, the better, since the tomato's roots can reach down as far as four feet.

A lower, harder mineral layer is called the geological base. It can be made up of rocks of any size, from pebbles to solid bedrock, and may contain little or no organic matter. If this layer is close to the surface, so that the soil atop it is very shallow, it may have to be dug out and replaced with loose soil. If it is bedrock, unfortunately, it is there to stay, unless the gardener counts a jackhammer among his tools.

Most gardeners will think their soil looks good enough to grow tomatoes, but others may be discouraged when they examine the soil profile. Those in certain parts of Florida, for example, may find there are no layers at all—only deep sand. Sand is so porous that water and fertilizer wash right through before roots have a chance to absorb them. (Extra watering won't do much to solve this problem—plants constantly draw on water held in the soil; a tomato plant will take up to about four pints of water on a sunny day.) Gardeners in other parts of the country may find that heavy clay in the subsoil has been compacted into an impenetrable layer called a hardpan. It will not absorb water at all; the soil above it will stay waterlogged and tomato plants will drown. Good soil for tomatoes will have more sand than clay; it will feel slightly gritty when rubbed between the fingers, yet will stick together when wet.

Perhaps the most important consideration is how well the soil drains. Fill the test hole with water and note how long the water takes to disappear. If the hole empties immediately, the soil is *too* sandy, and will not hold moisture and nutrients. If water is still in the hole after several hours, drainage is poor, possibly because the clay content is too high, or because a hardpan runs underneath. If the water drains from the hole in thirty minutes to an hour, drainage is ideal for growing tomatoes, and you need not worry about improving the soil's structure.

If, after examining the soil, you judge it to have a crumbly texture and good drainage, you can join the ranks of gardeners who claim that tomatoes simply grow themselves, and be reasonably sure that your garden will bring forth a bountiful harvest with or without much help from you. Few soils are so amenable, however, and most gardeners will find that their ground needs some work. Even good soil can be made better. And the better the soil, the higher the chances for enormous yields of flawless tomatoes on healthy, sturdy plants.

THE pH FACTOR

The sourness or sweetness of garden soil—the acidity or alkalinity—is measured on what is called the "pH scale." The numbers of backyard gardeners who have grown perfect tomatoes without the faintest notion of the pH of their soil, or without any effort to adjust it, are legion. Some gardeners are lucky enough to live on ground that naturally has the slight degree of sourness that tomatoes crave. Others dig enough organic matter into the soil that they unwittingly help to solve the problem: bacteria feeding on organic matter will usually adjust the soil's pH to a level that is right for vegetables.

Chances are, most gardeners can ignore pH and still come out with a bumper crop of tomatoes. But those who have been disappointed with their tomato crops in the past, who have despaired over curled, yellow plants with anemic fruits despite rigorous attention to watering and fertilizing, should consider that a too-high or too-low pH might be at fault. And raising or lowering pH is easy.

The tiny root hairs of plants can take up nutrients—nitrogen, phosphorus, potassium, and the thirteen other chemical elements they need—only when those nutrients are dissolved by natural acids in the soil. Some soils have more acids—are more sour, in common garden parlance—than others. In a garden with acid soil, phosphorus, for example, is easily dissolved and released to plant roots; in another garden, with alkaline soil, the phosphorus remains undissolved and therefore unavailable.

The letters pH stand for potential hydrogen, but it is hardly essential for home gardeners to understand the scientific context of acids and alkaloids. It is necessary to know, however, that the pH scale runs from 0 to 14, and that garden vegetables grow best in a range of 5 to 8. Soils with a pH on the lower end of the scale are acid; soils on the higher end are alkaline. Most garden soils will have a pH no lower than 4 and no higher than 8. Tomatoes prefer a slightly acid range—5¾ to 6½. If soil is more acid, needed potassium and magnesium will dissolve so quickly they will leach from the topsoil, out of the reach of the tomato plant's roots. Aluminum and iron will dissolve so well the plant will take up too much of them and risk being poisoned.

KITS AND COUNTY AGENTS

There is nothing complex or difficult about determining the pH of garden soil. But some ways are simpler than others.

Your county extension agent will do the work if you send him a soil sample. Some gardeners, however, may find this method too involved. It requires writing or calling the agent, or dropping by his office, to obtain a small soil bag, taking a few scoops of soil from various places in the garden and mixing them together, filling the bag and mailing it back to the agent. It is tested in a laboratory and a written report is sent back by mail. Some gardeners will be put off by having to make trips to the post office or extension agency, and by having to wait days—usually about ten—for the results of the test. One advantage is that the county extension service will not only report the pH of the soil, but also its mineral makeup—which tells you which fertilizers you need. Agents in some states charge a fee for this service.

Impatient gardeners probably will prefer to use a pH testing kit bought at a garden center or nursery. In this way, they can get an immediate pH reading. The kits, costing from $4 to $8, consist of test tubes partially filled with a solution. The gardener simply adds a little finely crumbled soil, lets it settle, and matches the color of the solution to the chart in the kit. This gives him a close enough reading to judge how he should adjust his soil, if at all.

Gardeners who live in areas that are known to have infertile soil may want to invest in a full-fledged soil testing kit—available for $12 to $20 at nurseries or garden centers—that will not only give a precise pH reading, but also will measure all sixteen chemical elements in the soil that plants need. Even gardeners who don't need such a kit may want to buy one. For one thing, it affords a chance to play with test tubes and solutions; for another, it gives the real enthusiast the satisfaction of knowing the chemical composition of his garden intimately—that, for example, the east side of the yard is richer in molybdenum than the west side.

Ross Smith, a print restorer in Baltimore, has a cheaper, quicker way of testing the pH of his soil: he uses the paper pH indicators designed to test the acidity of paper. (The paper strips, sold under the brand name Colorphast, are available in art supply stores. They are more precise than basic litmus paper, which simply turns red or blue. Gardeners should buy strips that measure the 4 to 7 pH range.)

Smith mixes a little crumbled soil, taken from 2 to 3 inches under the surface, with a few drops of distilled water, then dips the paper strip into the mixture for about 2 minutes. When the strip is removed, it will change to its final color after about 4 minutes. To obtain the pH reading, the color is then matched to a chart on the box that contains the strips.

DIGGING THE PLOT

Once you have determined whether your land is on the sweet side or the sour, and have a general idea of the soil profile, you can dig the tomato plot and amend it as necessary. The typical

131

12-X-3-foot plot, enough for six tomato plants, is small enough to tackle with a shovel. Gardeners who are preparing much larger plots, for several vegetables, may find it worth their while to rent a rotary tiller, which will take much of the work out of breaking up the soil.

Initial digging is best done in the fall, especially if the tomato plot is being carved out of a piece of lawn. Pierce the grass to a shovel's depth and turn it upside down so that soil and roots are exposed to the air. Exposure to the elements through winter and early spring will help break up rough clods and loosen the hold of grass roots.

Early in the spring, when you are ready to turn the bumpy, upturned sod into a finely textured, rich growing medium for tomatoes, put your knowledge of the soil to work. If you have judged your garden to be too sandy, add matter that will help it hold water and nutrients for the tomato plant's roots. The trick is to make the soil absorbent, and this is best accomplished with the addition of some kind of spongy organic matter—peat moss, manure, or compost. One large bag of peat moss (4 cu. ft.), dug in to a depth of 1 foot, is enough for a 12-X-3-foot plot. To make the peat workable, wet it well with a hose before spreading it evenly over the soil and forking it in.

Dehydrated animal manure is also a good material for building up sandy soil. Cow, horse, and chicken manure should be at least three months old before using, lest they burn the roots of newly set-out seedlings. (Chicken manure, the most likely to burn, should be thoroughly mixed with sand or straw before forking in; this dilutes its strength.) Spread the manure over the plot to a thickness of about 4 inches and work into the soil to a depth of 1 foot.

Compost, used by keen gardeners as an all-purpose soil improver, will also add structure to sand. Spread a 3-inch thickness of compost over slightly sandy soil and mix well; use up to 8 inches for soils that are almost pure sand.

If your problem is not too much sand, but too much clay, lighten the soil by digging in coarse sand. But be aware that for the sand to make a difference, there must be a lot of it. A heavy clay soil will need as much as one part sand to two parts soil. Spread the sand evenly over the plot and dig it in to a depth of

---

Jim Berry of Ocala, Florida, has a novel way of keeping water and nutrients from washing too quickly through his sandy soil. He buries a rotting pine log three feet under the ground and plants a row of tomatoes on top of it. As the soft log decays and breaks up over three or four years, it acts as a barrier and slows drainage, thus giving the tomato plants a chance to quench their thirst.

---

1 foot, turning the soil several times to mix well. And make sure the sand is coarse, since fine sand, such as that taken from the beach, will have little effect.

MAKING COMPOST

Compost will act as a conditioner—and a fertilizer, as well—for both sandy and clay soils, and has the advantage of recycling waste into the useful organic matter that gardeners are always seeking.

Organic matter is plant or animal material—anything that comes from a living source. Leaves, grass clippings, peat moss, and manure are the kinds most frequently used to condition the soil. They aerate it, and, as they rot, they break down into humus, the dark, sweet-smelling soil that is the best possible growing medium for tomatoes or any other crop.

Decaying organic matter contains acids that help dissolve the essential minerals the plants absorb. Fresh organic matter, however, may do little more than improve drainage; by the time it starts to break down to form humus, tomatoes may have produced their last fruits of summer. Fortunately, gardeners can "pre-rot" the organic matter they add to their soil simply by heaping garden and kitchen waste—leaves, grass clippings, straw, hay, fruit peels, egg shells, coffee grounds, even long-forgotten leftovers from the back of the refrigerator—into a corner of the garden to make a compost pile.

In the original Latin, compost means "bring together." In the garden, it means bringing organic materials together to commingle and rot. Piling them together with nitrogen-rich matter speeds up the process that turns leaves, stalks, rinds, and manures into fertile humus.

There is nothing mysterious or messy about compost piles. No special equipment is needed, and only a few loose rules apply. As long as organic matter is piled together, activated with nitrogen, watered, and aerated, it will "cook"—bacteria will feed on it, and in the process of digesting it, heat it up to 140° to 170°. The pile will even give off steam in the process; most weed seeds, fungal spores, and disease-causing bacteria will be killed. Composting garden waste is rather like boiling unpurified water to make it safe—the heat gets rid of just about anything undesirable.

Almost anything that is biodegradable can be turned into compost. Besides garden waste such as leaves, shrub prunings, grass clippings, weeds, faded flowers, and spent vegetables, you can add weathered sawdust, wood chips, shredded corn cobs, peanut shells, dog droppings, seaweed, sour milk, even human hair. And a few inorganic materials, such as marble dust or ground limestone, can be sprinkled in sparingly. Some things should be avoided, however: meat or animal fat will not degrade as quickly and as cleanly as vegetable matter; and human waste and cat droppings can harbor germs that may survive the heating process.

Tomatoes particularly like compost made from their own debris. At the end of the season, throw spent plants and rotting fruits onto the pile. (Diseased tomato plants should always be burned, however; if the compost pile does not heat sufficiently, the fungal spores and bacteria that caused the disease will winter over in the pile and be reintroduced to the garden in the spring.)

Since compost piles can be kept going indefinitely (some of the best are decades old), many gardeners prefer to construct some kind of permanent enclosure—a three-sided bin, a simple wooden frame, or a chicken wire fence. Other gardeners simply maintain a neat, unfenced pile, tapered as it rises, in a corner of the garden. In any case, to heat properly, the pile should be at least 4 to 5 square feet at the bottom.

For the material in the compost pile to "cook," it needs water, air, and nitrogen. The nitrogen, which gives bacteria something to feed on and thus initiates the decay process, can be provided by horse or chicken manure. Both should be very dry before

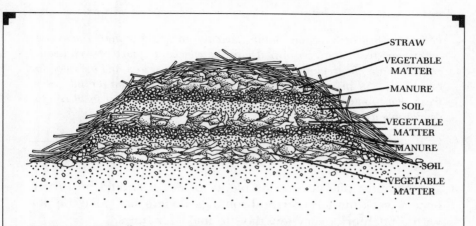

STRAW

VEGETABLE MATTER

MANURE

SOIL

VEGETABLE MATTER

MANURE

SOIL

VEGETABLE MATTER

Start a compost pile by spreading a 4- to 6-inch layer of leaves, kitchen waste, or other vegetable matter over 4 to 5 square feet of bare earth, gravel, or straw. Top with an inch or two of soil, then with a layer of animal manure or other nitrogen-rich material. Repeat the layers as vegetable matter becomes available, and keep the pile topped with straw to shed rainwater.

adding to the pile, and chicken manure should be mixed with chopped straw. Alfalfa meal, blood meal, cottonseed meal, or chemical nitrogen fertilizers are also good "starters." If using animal manure, add it at the rate of one part manure to five parts plant material; if using meals or chemical fertilizers, sprinkle on at the rate of one cup for every 10 square feet of level pile surface.

Randy Drinkard of Marietta, Georgia, turns autumn leaves into rich compost by storing them over the winter in large black plastic leaf bags. After filling a bag with leaves, Drinkard puts in a small shovelful of soil and a handful of 10-10-10 fertilizer as an activator. He saturates the leaves with water, ties the bag and bounces it on the ground a few times to mix the ingredients. He then stores the bags in a sunny place, so that they will absorb the heat of the sun. By spring, the leaves have rotted to form fertile humus.

Polly Marshall of Bowman, South Carolina, makes a compost activator, or starter, from the compost itself. She places 1 quart of well-heated compost in a cheesecloth bag and leaves it to steep in a pail of water for one week. The strong, dark "tea" it brews is poured onto the pile when fresh plant material is added. It also makes a good liquid fertilizer for garden plants.

A compost pile also needs plenty of air. It should be turned once a week for aeration, and some gardeners also poke the pile with a pitchfork every few days to make air holes.

Dave Westmeier of Berea, Kentucky, finds it hard to aerate his large, 20-year-old compost pile by turning, so he creates "ventilation stacks" by adding old sunflower stalks to the mixture when he adds leaves and other material. He layers the stalks into the pile every foot or so, with their ends protruding. The centers of the stalks rot out, and the hollow poles that remain channel air into the center of the pile.

ADJUSTING pH

Once the soil has been conditioned, consider the results of your soil test (page 130), and adjust the sweetness or sourness of your tomato plot accordingly. A soil test in most parts of North America will show that garden soil is too acid—that is, below the 5¾ to 6½ range that tomatoes prefer. In some western states, and in isolated pockets in the rest of the continent, the soil may be too alkaline. If you live in a state where the soil is said to be generally acid or alkaline, do not be secure in the assumption that your garden fits the norm; many gardens do not, and only a soil test can tell you whether yours is one of them.

If, indeed, a pH test shows your soil is too acid for tomatoes and needs to be sweetened, plain ground limestone, hydrated lime, or dolomitic limestone will do the trick. It also helps fertilize the soil, providing calcium, an important nutrient for the prevention of blossom end rot of tomatoes. (Gardeners who know their soil to be low in magnesium should use dolomitic limestone, sold as dolomite, which provides it.)

Lime can take up to three or four months to change the pH level of the soil, so the best time to apply it is in the fall, well before next spring's tomatoes are set out in the garden. Mix it in thoroughly with the top 6 to 12 inches of soil, since merely sprinkling it on the surface will have little effect.

The sand or clay content of the soil determines how much lime should be added. In gardens with soil on the sandy side, dig in 2½ pounds of ground limestone for every 100 square feet of soil; if using dolomite, dig in 2 pounds; if hydrated lime, use 1½ pounds. Heavy, clayey soils need more: 6 pounds of ground limestone for every 100 square feet; 5½ pounds of dolomite; or 3½ pounds of hydrated lime. These additions should raise the pH by a full point, usually enough to bring soil into the 5¾ to 6½ point range for tomatoes.

Instead of lime, Katherine Bowlin of Nashville, Tennessee, uses the wood ashes from winter fires to sweeten her garden soil. When she cleans her wood-burning fireplace, she stores the ashes in a can in the cellar. In late winter, she sprinkles a 3-gallon bucketful of the ashes evenly over her 10-X-10-foot tomato patch when she prepares the soil for planting, and digs them into the top 3 inches of soil. The ashes not only raise the pH of the soil, but supply valuable potassium.

Some gardeners, especially those in the West, may need to make their soil more acid rather than alkaline. This can be accomplished by adding iron sulphate, aluminum sulphate, or plain sulphur. For sandy soils, add 3 pounds of iron sulphate, 2½ pounds of aluminum sulphate, or 1 pound of sulphur for every 100 square feet. For heavy, clayey soil, add 7 pounds of iron sulphate, 6 pounds of aluminum sulphate, or 2 pounds of plain sulphur. Dig evenly into the top 6 to 12 inches of the soil.

Russell Harrison of Redding, California, uses the time-honored method of tilling oak leaves into the garden to increase its acidity. If a yearly October pH test shows the soil needs acid, Harrison tills in the fallen leaves from the red oaks surrounding his home, using ten 24-gallon bags of leaves for the entire 900-square-foot plot. By spring, the pH has been lowered by about one point, and the soil is sufficiently sour for tomato plants.

FERTILIZING THE SOIL

Organic matter, lime, and sulphur will go a long way toward enriching the soil, improving its structure, or adjusting its acidity, but additional fertilizer is usually needed to nudge tomato plants—especially F_1 hybrids—into performing their best. Each plant's eventual yield will be increased if you sidedress them with fertilizer every couple of weeks during the growing season (page 157), but you also need to dig fertilizer into the plot when you prepare it for planting.

Complete fertilizers, the ones referred to numerically—10-10-10 or 5-10-10, for example—are the most commonly used. They are synthetic rather than "organic," and for this reason have fallen from favor with those gardeners who are reluctant to put anything "unnatural" into the soil that grows their food. In line with the swift pendulum swing that has made seemingly artificial supermarket tomatoes unacceptable nowadays, people increasingly want things to be natural, including the fertilizers with which they feed their plants. They have the idea that so-called organic methods—those that avoid the use of any chemical fertilizers or pesticides—will result in larger, more nutritious, and better tasting tomatoes. There is indeed something wholesome and attractive about nurturing tomatoes only with the inherent products of the earth instead of with dusty, factory-made fertilizers that reek of chemicals. In truth, however, scientific test after scientific test has shown that there is no ·appreciable difference between organically and inorganically grown vegetables, including tomatoes, in terms of yield, size, nutrition, or taste.

This is not to suggest that pure organic gardening is not

worthwhile. While it may not be feasible on a large commercial scale, it works well in the home garden—especially in the northern states and Canada, where garden pests are more easily controlled. Most gardeners try a little of everything these days, using some organic techniques, such as maintaining a compost pile, yet adding man-made fertilizers or mild pesticides as needed.

Complete fertilizers supply nitrogen, phosphorus, and potassium, the three elements that plants need most. The numbers with which the fertilizers are labeled refer to the percentages of these three elements, whose chemical symbols are N, P, and K. Thus, in a 100-pound bag of 10-10-10, there are ten pounds of pure nitrogen, ten pounds of pure phosphorus, and ten pounds of pure potassium; the remaining seventy pounds are fillers that dilute the elements to a form that can be handled by plants.

Nitrogen, phosphorus, and potassium in chemical fertilizers are no different from that in organic matter. Plants absorb nutrients only after the nutrients have been broken down into ionic form, and an ion from a synthetic fertilizer cannot be distinguished—by a scientist or by a tomato plant—from an ion obtained from an organic source.

Tomatoes need more phosphorus and potassium than nitrogen. Too much nitrogen will result in leaf growth at the expense of fruiting. A 10-10-10 fertilizer, therefore, is not as suitable as a 5-10-10, 6-8-8 or 8-16-16. (Numbers on fertilizer labels are always in the same order, with the nitrogen content first, phosphorus second, and potassium third.) Mix complete fertilizers into the top 6 inches of soil when you prepare it for planting; instructions on the bag will tell you how much to apply.

Superphosphate is a commercial fertilizer containing only acid-treated rock phosphate; the acid helps it break down so that it can be absorbed by plants. Superphosphate bags are labeled 0-20-0, since they contain no nitrogen or potassium. Tomatoes are always hungrier for phosphorus than they are for other nutrients, so many gardeners spread superphosphate over the tomato plot—5 pounds for every 100 square feet will last two or three years—and dig it in before setting out plants in the spring.

Bone meal is the organic gardener's main source of phosphorus. Other readily available sources are cottonseed meal and dried blood, all sold at garden centers and hardware stores.

Myrtle Dohn of Reynoldsburg, Ohio, has an effective, if unusual, way of supplying phosphorus to her tomato plants. Her technique is not for the squeamish, however: she obtains fresh beef blood from the local butcher and pours a quart into each planting hole. The blood apparently supplies not only phosphorus, but other nutrients as well, since Mrs. Dohn's tomato plants inevitably outproduce those of her neighbors, even though she adds no more fertilizer for the rest of the season.

Animal manure adds body to sandy soil and lightens heavy ones (page 132), but is also a rich source of nitrogen. Cow and hog manures have less nitrogen than horse, chicken, rabbit, and sheep manures, but all are effective if dug into the soil when they have aged for at least three months. (To be on the safe side, mix chicken manure with chopped straw; of all manures, it is the most powerful, and the most likely to burn plant roots.) Manure also supplies some phosphorus and potassium.

Many gardeners complain that manure contains weed seeds that later sprout lavishly in the tomato patch. The surest way to prevent the problem is to add the manure to a compost pile before using, making sure the pile heats to at least 160° in the center; the heat will kill weed seeds. Apply manure by spreading a 3-inch layer over the soil and digging it in, mixing well.

Wood ashes provide potassium, and also make soil more alkaline (page 129). An alternate, though expensive, source of potassium is greensand marl, mineral matter that comes from the ocean floor. It should be applied at the rate of 5 pounds for every 100 square feet of soil.

RAISED BEDS

When digging the soil and amending it, you may find it worth your while to do a little extra work to raise the garden bed above ground level. If a drainage test (page 127) has shown that your soil stays waterlogged, raising the planting bed will keep tomato plants from drowning. Not only will the soil drain freely, but it will also warm up more quickly than surrounding

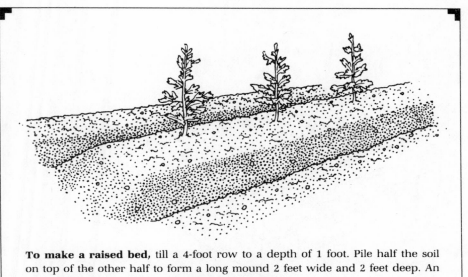

To make a raised bed, till a 4-foot row to a depth of 1 foot. Pile half the soil on top of the other half to form a long mound 2 feet wide and 2 feet deep. An occasional raking will maintain the bed's shape.

ground in the spring, a favor to heat-loving tomatoes. Northern gardeners can add as much as two weeks to the growing season by setting out seedlings in the warm soil of a raised bed.

Although many gardeners assume that raised-bed gardening involves building a frame of some sort with wooden planks or railroad ties, it can be as simple as mounding up the soil into the kind of hill on which cucumbers are planted. Even raising enough ground for a long row of tomatoes, or a complete vegetable garden, is easy. Just dig or till the entire plot and mark off rows with stakes and string (2 feet is wide enough for a raised row of tomatoes), with walkways in between. Rake all of the tilled soil from the walkways onto the rows to form ridges. This means that if you have dug your soil to a depth of 1 foot, the raised planting beds will have a 2-foot depth of loose, friable soil. And the lower walkways will give you access to the plants with less bending and stooping.

Once the soil has been raked into ridges, smooth the top of

141

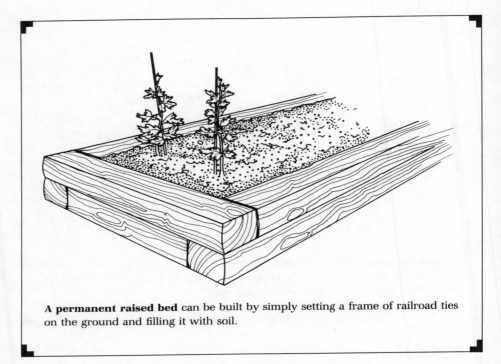

A permanent raised bed can be built by simply setting a frame of railroad ties on the ground and filling it with soil.

each row with a rake until it is level. The bed is now ready for planting. An occasional raking will keep it in shape through the growing season.

DEEP-HOLE PLANTING

As an alternative to digging a whole row or plot, some gardeners prefer to prepare a wide, deep planting hole for each tomato plant, mixing the soil in the hole with heavy helpings of organic and mineral matter. Soil can be dug more deeply than would be practical in a whole plot, and tomatoes thrive when their roots can snake down unimpeded into enriched earth. Most will grow so well in a deep hole that they can be virtually ignored—except for the occasional watering—for the rest of the growing season. Placing a wire cage (page 161) over the hole eliminates staking and pruning later in the summer, and a mulch of straw around the cage conserves moisture and keeps weeds from sprouting. Thus can tomatoes "grow themselves."

Tomato roots sometimes reach a length of 4 feet, but a hole that is 2½ to 3 feet deep is sufficient to guarantee that tomatoes will perform better than those grown in a plot tilled to the usual depth of 1 foot. And instead of simply putting the freshly dug soil back into the hole, the gardener can concoct his own recipe—based on his knowledge of the soil—to create the richest possible growing medium. In all cases, plenty of organic matter is needed, and certain minerals and fertilizers will help correct deficiencies that may have been revealed by a soil test (page 130). Almost anything organic, from fish heads to egg shells, can go into a deep hole, so that minicompost piles are formed under growing plants. Trying a different combination of materials in each hole for a couple of summers will help you develop the one formula that will nurture your tomato plants better than any other.

John Davidson of Austin, Texas, uses the waste from last year's corn crop to give his tomatoes a boost. He places about 5 inches of corn cobs and shucks at the bottom of a 2-foot-deep hole, covers them with cow manure, then shovels soil back into the hole without mixing the three ingredients. The cobs and shucks act as a sponge to conserve moisture in the hot, dry Texas summer.

Lou Peneguy of Atlanta, Georgia, has found that placing a few fresh banana peels at the bottom of the planting hole makes his tomato plants noticeably more vigorous. He simply puts 3 or 4 peels at the bottom of each hole and shovels a mixture of leaves, manure, and soil on top. The peels provide a kind of time-release fertilizer, leaching potassium and trace minerals into the soil as they decay.

SEED

It is undeniably easier to wait until spring and buy tomato seedlings at a nursery than it is to buy seed in late winter and germinate it under lights. Buying seedlings is also more expensive,

Collie Moon of Annistown, Georgia, has formulated a soil recipe that yields an enormous crop of tomatoes every year. Sometimes the fruits are surprisingly large: in 1981, one of Moon's Supersteak plants produced a 3-pound-4-ounce tomato that won a Georgia-wide contest for the biggest tomato grown by a home gardener.

For each tomato plant, Moon digs a hole about 3 feet wide and 2½ feet deep. In Moon's part of the country, this means digging out at least 1½ feet of red clay subsoil. He tosses in a 3-gallon bucket of rotting leaves, followed by a bucketful of well-aged cow manure or chicken manure mixed with straw. Next come a quart of cottonseed meal and a quart of bone meal, measured with a milk carton. Since Moon's soil is on the sour side, he adds about half a shovelful of lime, and finally, a heaping shovelful of granite dust obtained from a nearby quarry. (Granite dust, also called rock sand, is actually a fine gravel. It is rich in trace minerals, but whether those minerals actually leach into the soil and are taken up by the plant is open to question. The main purpose of the "dust," the pieces of which range up to ⅛ inch, is to loosen the soil.)

Moon then shovels soil—minus any clay—back into the hole, and mixes all of the ingredients with a pitchfork or shovel. Then he tamps down the soil so that the planting hole is slightly below ground level; this allows it to collect rainwater efficiently.

After Moon transplants a seedling to the prepared hole, he surrounds it with a wire cage, and, after a few weeks, mulches the plant with 4 to 6 inches of straw. His tomatoes, except for a weekly watering, then take care of themselves. Their roots grow effortlessly in the fertile, friable mixture and bring forth large, perfect fruits until the first frost.

---

but most gardeners think it well worth the extra cost, since it frees them from the chore of starting tomatoes indoors. (Gardeners in warm climates have the luxury of sowing seed directly in the garden, but those in areas with short growing seasons almost always have to set out seedlings, whether grown at home or bought at the nursery, as soon as possible after the danger of frost has passed.)

Why, then, would anyone choose to bother with seed instead of buying seedlings? There are, in fact, several advantages. The main one is that the gardener can avail himself of an incredibly wide range of varieties. Once he has acquainted himself with

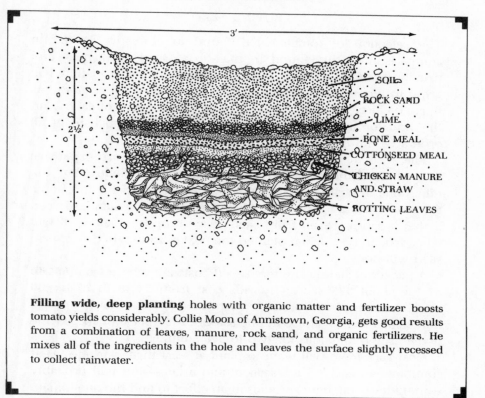

Filling wide, deep planting holes with organic matter and fertilizer boosts tomato yields considerably. Collie Moon of Annistown, Georgia, gets good results from a combination of leaves, manure, rock sand, and organic fertilizers. He mixes all of the ingredients in the hole and leaves the surface slightly recessed to collect rainwater.

the varieties it is possible to grow, he may not want to limit himself to the selection found at the nursery, hardware store, or supermarket. The seedlings sold there will suit the local climate, no doubt, but may be stocked simply because the local folks are used to them, or because the retailer is not aware of better choices. In hardware stores or supermarkets, and even some nurseries, seedlings may remain on shelves so long that they become leggy and weak, and may never thrive when transplanted. They may even be mislabeled, so that the gardener who expects a Better Boy ends up with an Early Girl. Buying seed yourself removes all doubt about what you are getting. And germinating it yourself, in sterilized soil, guarantees that your seedlings will be free of the diseases that are sometimes introduced into the garden by store-bought plants.

BUYING SEED

The selection of tomato seed found in stores is limited in comparison to that found in mail-order seed catalogues, and usually includes only the "widely adapted" varieties that will grow in almost any climate zone. Catalogues offer as many as thirty or forty varieties, with a detailed description of each. And the wait for mail-order seed is not long: seedsmen usually fill orders within two weeks of receiving them.

The seed of F_1 hybrids is 20 to 50 percent more expensive than that of the standard (open-pollinated) varieties, but the difference in cost is negligible when buying seed for the home garden. The only economic advantage in buying standard seed is that it can be saved from each year's crop (see below); it will come true to type when planted the next season, while F_1 hybrid seed will not.

A packet of standard tomato seed costs anywhere from 65¢ to $1 for about 125 seeds; hybrids cost from $1 to $1.25 for 30 seeds. In both cases, there are more seeds per packet than most home gardeners will use in one season. A $1 packet can provide home-grown tomatoes for two to five summers.

Most gardeners will want to plant at least three varieties each summer—an early, a midseason, and a late—and will probably want to try some new varieties in an effort to find the one tomato that best suits their garden conditions and their taste. Six packets of seed, each of a different variety, should be sufficient for even the most experimental of gardeners, and will cost only $5 to $8. This is obviously good value for the money, considering the seeds will last for up to five years and will supply more tomatoes than the average family can use. It is especially good value when one considers that the price of one packet of seed is about the same as the cost of three inferior supermarket tomatoes at the height of summer.

SAVING SEED FROM YOUR OWN TOMATOES

Gardeners who have settled on a few favorite nonhybrid varieties need never buy seed again. Instead, they can plant the seeds of last year's tomatoes. (Standard, or open-pollinated, varieties come true to type, while F_1 hybrids revert in the second generation to their numerous ancestral strains, few of which may be edible.) Saving your own seed is not only cheaper, but allows you to select the best tomatoes in your garden and perpetuate their characteristics. By doing so, you can gradually develop your own improved strain.

Saving seed is simple. First, find the plant that has performed best all season—the one that has the healthiest leaf growth and has produced a good yield of uniformly sized fruits. When the largest, best-shaped tomatoes on the plant are dead ripe, cut them open and scrape out the center pulp, pulverizing it with your fingers. Place it in a glass jar for about 48 hours, to allow the acid in the pulp to kill any bacteria that may be carried on the seeds. When the seeds have floated to the top, strain them in a sieve under running water and spread them out on paper towels to dry in a fairly warm place—70° to 75° is ideal. When the seeds are completely dry, put them in an envelope—not in a lidded jar—so that they will be exposed to oxygen during storage. Write the variety name and date on the envelope and store in a cool place. The seeds should remain viable for up to five years.

GERMINATING SEED INDOORS

Although the apparatus for starting seed indoors can be as basic as a paper cup and a little dirt, innumerable products are sold in garden stores to make the job as neat and effortless as possible. Some gardeners think such equipment—including expanding peat pellets, special potting mixes and even electrically warmed seed-starting kits—is unnecessary, since healthy tomato seed will inevitably germinate within 7 days in any medium if given enough heat, light, and water.

Whether germinating seed with expensive equipment or with whatever is at hand, the procedure should begin 6 to 12 weeks before you plan to set seedlings out in the garden. In the northern states and Canada, this generally means starting some time in March; gardeners in the South and Southwest might start as early as January or February to ensure an extremely early harvest at the beginning of their long growing season.

Containers can be a paper cup, sawed-off milk carton or small flower pot—anything is suitable, as long as it is at least 3 inches in diameter and has drainage holes in the bottom. Wooden or plastic flats, big enough for starting at least a dozen plants, are also used, although the seedlings need to be repotted in individual containers for a few weeks before they are transplanted to the garden. Individual peat pots, cubes, and expanding pellets are available for germinating seeds, and have the advantage of being placed directly into the planting hole without removing the seedling and disturbing its roots; when covered with dirt, the peat containers will quickly rot away.

Soil for germinating seed can be taken from your own garden. To prevent damping off (page 187), it is essential to sterilize garden soil by spreading it on a cookie sheet or in a large shallow dish and heating it in an oven at 350° for at least 30 minutes.

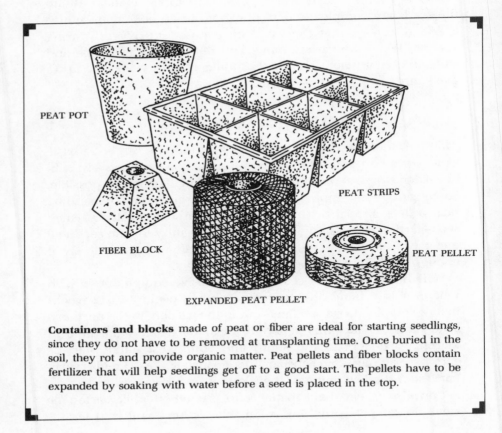

PEAT POT

PEAT STRIPS

FIBER BLOCK

PEAT PELLET

EXPANDED PEAT PELLET

Containers and blocks made of peat or fiber are ideal for starting seedlings, since they do not have to be removed at transplanting time. Once buried in the soil, they rot and provide organic matter. Peat pellets and fiber blocks contain fertilizer that will help seedlings get off to a good start. The pellets have to be expanded by soaking with water before a seed is placed in the top.

Most garden soils should be mixed with vermiculite or perlite (minerals commonly used to aerate potting soils) or with peat moss to attain the lightness that helps newly sprouted seeds to thrive.

Commercial potting soils are presterilized, and are designed to hold just the right amount of moisture. They usually contain peat, as well as vermiculite, perlite, and other minerals. A 3-cubic-foot bag, enough to last most seed-starters for several years, costs around $12.

Light is better provided by fluorescent bulbs than by sunny windows. For seeds to germinate, they need a fairly even temperature—and a windowsill will probably be too warm during the day and too cold at night. Containers should be kept under lights for about 16 hours each day, and need at least 8 hours of total darkness. When seeds have germinated, the plants should be kept about 2 inches away from the fluorescent bulb, lest they grow leggy from reaching for the light. Contrary to some gardeners' beliefs, special ultraviolet plant lights, of the kind used for house plants, are not necessary for starting tomato seeds indoors. Regular fluorescent lights work fine.

Heat is needed for tomato seeds to germinate. Temperature in the immediate vicinity of the containers should be between 75° and 80°. After germination, seedlings grow best at temperatures between 75° and 80° during the day and between 60° and 65° at night.

When you are ready to plant seeds, fill your pots, cups, or flats with sterilized soil or potting mix and moisten them well. Place three seeds in a pot or cup, or space seeds ½ inch apart in a flat. Press the seeds about ¼ inch into the soil and gently scrape soil over them. Put the containers in a well-lit, warm place. A sheet of plastic wrap laid over pots and cups will keep them moist enough so that they need no further watering. Flats can be enclosed in a large plastic bag to conserve moisture.

The seeds should sprout in 5 to 7 days. When they have developed a second set of leaves, thin the three seedlings in pots or cups to the strongest one; thin those in flats to no more than 12. Seedlings grown in flats will have to be transplanted to separate containers before they are large enough for transplanting. When the third set of leaves has formed, lift the tiny seedlings from the flat and place each in a 3-inch pot or cup, burying the

Jim Dershem of Stow, Ohio, pots his seedlings indoors in progressively larger household containers—from small paper cups to half-gallon milk cartons—before setting them out in the garden. This enables his plants to develop a strong root system that speeds their growth.

Dershem germinates seed in a 3-ounce Dixie cup with a few small drainage holes punched in the bottom. When the seedlings are about 5 inches tall, he tears the cup away from the root ball, which is now firm and matted. He places the seedling in a 14-ounce paper cup so that the leaves extend at least 2 inches over the top of the cup. He then fills the cup almost to the top with potting soil; at least half of the stem is buried. (The buried portion of the stem develops new roots as the plant grows; leaves should be stripped from the lower portion of the stem before it is buried.)

When the seedling has grown 6 to 7 inches tall, Dershem removes the seedling from the cup and places it in a half-gallon milk carton. He fills the carton almost to the top with potting soil, again burying part of the stem.

When he is ready to transplant to the garden, Dershem carefully tears the container away from the matted root ball. (The plant has become root-bound three times by this point, and the root system has grown stronger and denser with each potting.) He digs a 6-inch furrow in the garden and lays the root ball on its side, bending the stem and holding it erect while covering the roots and the lower part of the stem with soil— a method known as trench planting. Because of its extraordinarily strong roots, the plant continues its growth without the usual transplant shock that affects most newly set-out seedlings.

stem in soil to within ½ inch of the bottom leaves. Let the plants grow in a warm, well-lit place until they are 6 inches tall. Keep the soil moist, but not waterlogged.

Gardeners who start their seeds as early as 10 to 12 weeks before transplanting find it worthwhile to repot seedlings two or three times.

PLANTING

As the weather grows warmer, you should "harden" home-grown seedlings to prepare them for transplanting. This simply involves gradually acclimating them to the outdoors by setting

Trench planting makes it easy to plant leggy seedlings or those that have been grown in tall containers such as half-gallon milk cartons. Lay the root ball on its side in an elongated hole dug to a depth of at least 6 inches. Hold the stem erect while covering the root ball and the lower part of the stem with soil.

them outside for a while each day. After about 2 weeks, they will have become accustomed to the different humidity, wind, brighter light, and lower temperature of their new environment. Seedlings that are not hardened before transplanting will survive if they are protected on chilly nights (page 153), but will get such a slow start that your eventual harvest of tomatoes will be delayed.

Tomato seedlings should be exposed to the outdoors in the

same way people go about getting a suntan—by starting with a few minutes of exposure and gradually lengthening the time each day. On the first day, place the plants in a well-lit, protected spot for 20 or 30 minutes before taking them back inside. Be sure not to place them in direct sunlight for the first few days; it may sunburn their tender leaves. Leave them out for progressively longer times each day, so that within 2 weeks they can stand a full day outdoors. Near the end of the 2 weeks, they can be left out overnight if the temperature falls no lower than 45°

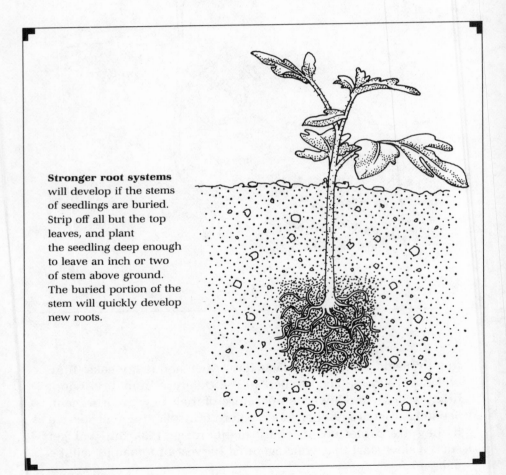

Stronger root systems will develop if the stems of seedlings are buried. Strip off all but the top leaves, and plant the seedling deep enough to leave an inch or two of stem above ground. The buried portion of the stem will quickly develop new roots.

to 50°. It also helps to toughen up seedlings not to water them much during the hardening period, nor should you fertilize them.

Nursery-bought seedlings have already started the hardening process in some cases by virtue of being displayed on outdoor racks. If they look large and sturdy, they can be transplanted into the garden without delay.

When all danger of frost has passed, and you have prepared either a garden plot (page 131) or custom-made deep holes (page 142), it is time to plant. Water the seedling while it is still in its pot, so that its soil will cling to the root ball. Then strip the lower leaves from the stem, leaving only the top two or three leaves. Remove the plant from its container, trying not to disturb the roots. (Peat containers do not need to be removed, since they will quickly rot once planted.) With a spade, lift out about 1 foot of soil and place the seedling in the hole so that only the top leaves and about an inch of stem are above ground level. Spade the soil back into the hole and firm it with your hands. This kind of deep planting, which buries most of the stem, will strengthen the plant, since the stem develops new roots under the soil. If the seedling is tall and leggy, it may have to be planted on its side in a trench (illustration, page 151).

Fertilizing home-grown seedlings at planting time can help them get off to a better start. High-phosphorus starter solution, sold at garden stores, can be poured into the planting hole, and will feed the plant until it begins to set fruit. In the case of store-bought seedlings, it is safer not to fertilize for a few weeks, lest you overdose them: they may have been recently fertilized at the store.

PROTECTING TRANSPLANTS FROM COLD

Even though the danger of frost should have passed by the time tomatoes are transplanted to the garden, the weather still may not be warm enough to enable the plants to thrive. Without some kind of insulation from chilly winds and low night temperatures, tomato plants will remain almost dormant until day temperatures reach the 70s; and they cannot tolerate prolonged temperatures below 50°.

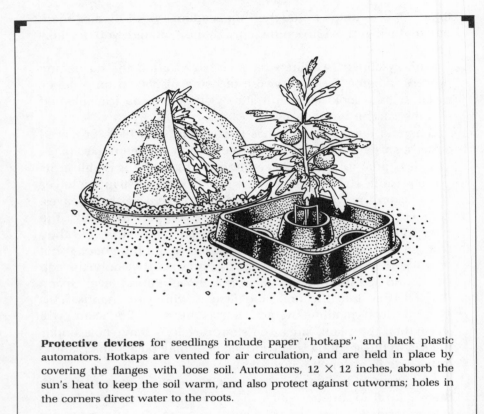

Protective devices for seedlings include paper "hotkaps" and black plastic automators. Hotkaps are vented for air circulation, and are held in place by covering the flanges with loose soil. Automators, 12 × 12 inches, absorb the sun's heat to keep the soil warm, and also protect against cutworms; holes in the corners direct water to the roots.

Protective coverings are usually necessary for at least a few days at the beginning of the growing season in all but the warmest climate zones. In the northern states and Canada, they are often essential for weeks. A number of commercial products are sold, including paper domes called hotkaps, and nifty plastic devices called tomato automators, which absorb the heat of the sun to warm the soil. Kitchen pots or inverted flowerpots will work as well, however, on isolated nights when seedlings run the risk of being nipped by frost. Other commonly used devices are gallon-size plastic milk jugs with the bottoms cut out (they should be pushed deeply into the soil so that they are not knocked over by wind) and large cardboard boxes held in place by heavy rocks. Some gardeners bend a sheet of corrugated fiberglass over a row of seedlings to act as a minigreenhouse.

Tomatoes are difficult to grow on the cold Olympic Peninsula of Washington State, but Charles E. Lockhart gets a good crop of Early Girls every year on his farm at Chimacum by warming the soil with a newspaper mulch and stacking tires around the growing tomato plants. Lockhart lays several thicknesses of newspapers over his entire plot, setting them right up to the stems of the newly set-out seedlings. Then he places a tire around each plant. As the plants grow taller, he adds a second and third tire. As a result, he is able to harvest ripe tomatoes in August—unusually early for his part of the country. While Lockhart's method may not be practical for the average suburban gardener, who may justifiably fear that a garden full of old tires and newspapers will be an eyesore, it offers an easy, workable solution for determined tomato growers in extremely cold climates.

Growing plants in a wire cage (page 161) makes insulating even easier, since material can be wrapped around the bottom part of the cage as protection from cold wind. A strip of black roofing paper, 12 to 18 inches wide, will also trap heat. Researchers in Texas have found that black roofing paper can warm seedlings so effectively that they mature earlier and produce up to 50 percent more fruits—a big payoff for such a simple procedure.

On cold days in early spring, Ralph Iannazone, a fruit dealer in Elizabeth, New Jersey, covers young tomato plants with slatted bushel baskets. As well as protecting seedlings from cold winds, the baskets let in some light. At night, Iannazone covers the baskets with a sheet of black plastic, anchored by rocks, for extra protection.

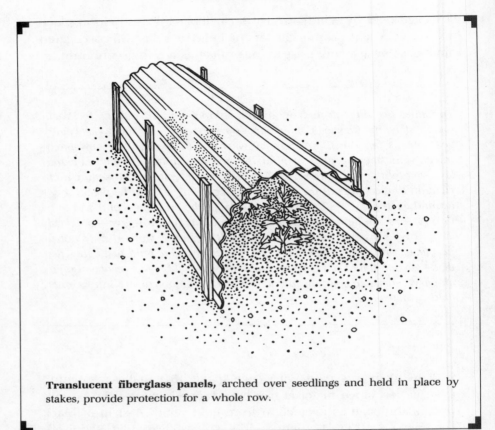

Translucent fiberglass panels, arched over seedlings and held in place by stakes, provide protection for a whole row.

CARING FOR PLANTS

The more carefully you have prepared the soil for tomatoes, and the more attention you have paid to selecting varieties that are adapted to your particular set of garden conditions, the less work you will have to do for the rest of the growing season. Watering is necessary, of course, and an occasional sidedressing of high-phosphorus fertilizer will help fruits maintain their size through the summer. But daily care is not essential, especially if you are growing varieties that do not need to be staked and pruned.

Watering is required almost every day if new seedlings are wilting, but can be gradually decreased as the seedlings grow stronger. By the time plants are well established, watering every 5 or 6 days should be sufficient; and if you surround plants with a moisture-saving mulch, water need be applied only every 10 days to 2 weeks, depending on how often it rains. In any case, an occasional slow, deep watering is preferable to a perfunctory daily sprinkling. Water that penetrates only a few inches into the soil makes tomato roots grow toward the surface to seek moisture; then, when the soil gets hot and dry, the barely protected roots die off. To water slowly and deeply, use the poor man's version of the sophisticated drip irrigation systems of commercial growers: a flat plastic sprinkling hose, the kind with holes in the side. Lay the hose alongside a row of tomato plants and turn it on so that it barely trickles. After a few hours, water will have soaked deep into the root zone without wetting the tomato plant's leaves (an advantage, since wet leaves encourage the spread of fungus diseases). If you repeat this trickle irrigation at regular intervals every few days, tomatoes will also be less

---

Vance Muse of Houston, Texas, waters his tomatoes through a juice can reservoir placed next to each plant. At transplanting time, Muse digs a small hole next to each newly set-out seedling and sinks a 46-ounce juice can into it. The can, with the top cut out and a few large holes punched in the bottom, is left in place for the rest of the season, efficiently channeling water to the tomato plant's root zone. Muse adds a cup of 5-10-10 fertilizer to the can every 5 or 6 weeks, so that nutrients trickle out with each watering.

---

likely to develop the blossom end rot (page 169) that can be caused by uneven moisture.

Sidedressing tomatoes every few weeks with fertilizer boosts yields and helps to keep fruits from decreasing in size as the season progresses. Such feeding is especially important for hybrid varieties, which depend on heavy helpings of fertilizer for their big crops. Be sure the plants have started to set fruit before you sidedress them; otherwise, you may encourage leaf growth at the expense of fruiting. When the first tomatoes on the plant have reached about 1 inch in diameter, sprinkle ⅓ cup of a high-phosphorus complete fertilizer—a 5-10-10 or 6-12-12, for example—in a circle around the stem of the plant and water it in well. (Do not let the fertilizer touch the stem.) Repeat the application every 4 to 6 weeks until the last tomatoes have been picked from the vine.

Hylan Davis of Corsicana, Texas, finds he gets better results from sidedressing with superphosphate, instead of a complete fertilizer, on his heavy, nitrogen-rich black soil. Once his tomato plants have started to bloom, he lightly scratches ¼ cup of superphosphate into the soil and waters it well. A repeat application every 3 weeks, Davis says, almost doubles the yields of the Better Boys, Super Fantastics, and Floramericas he grows every year.

Other good fertilizers for sidedressing include fish emulsion, dried blood, and well-aged poultry manure, all of which are high in phosphorus.

Mulching tomato plants—blanketing the soil around them with 4 or 5 inches of leaves, pine straw, bark, or even a few thicknesses of old newspapers—goes a long way toward improving the tomato crop. The benefits are many: Mulch softens the impact of rainwater and allows it to seep gently into the soil; it reduces evaporation, keeping the soil moisture even; it prevents soil erosion; it suppresses weeds; and it can be dug into the soil at the end of the season to provide organic matter for next year's plants. When placed around unstaked plants, mulch provides a buffer between fruits and the soil-borne organisms that may cause them to rot. And in hot climates, where the soil can grow too warm even for tomato plants, mulch keeps the roots cool.

Almost anything can be used as a mulch—leaves, compost,

pine straw, wood chips, peat moss, nut shells, aged manure, even seaweed. Some organic mulches will decompose as the season progresses, so replenish them as necessary.

Because mulch insulates the soil, do not spread it around seedlings for the first 4 or 5 weeks; mulching too early will keep the soil too cool and get the plants off to a slow start.

Gardeners in hot climates should not use black plastic as a mulch, since it will heat up soil to a degree that even tomatoes find uncomfortable. The purpose of mulching in the South and Southwest is to keep the soil fairly cool.

Some mulches make the soil warmer instead of cooler. John Zervos of Bethlehem, Pennsylvania, uses black plastic sheeting as a mulch, a practice common among northern gardeners. The plastic traps enough heat in the soil to get tomatoes off to a much earlier start. Zervos places a 2-foot-wide sheet of black plastic the length of his 12-foot tomato bed, and plants seedlings through slits cut every 18 inches. He anchors the plastic by shoveling dirt over the edges, and punches a few small holes into the plastic to admit water. The blanket of plastic keeps the soil permanently warm and moist.

Alice Gordon of Tulsa, Oklahoma, places sheets of aluminum foil around the base of her tomato plants on hot days to reflect the sun and keep the plants' roots cool. Other reflective materials will serve the same purpose.

Weeding is rarely necessary when a thick layer of mulch is laid down. But gardeners who grow their plants without the benefit of mulch should pick weeds while the plants are still small, before they have a chance to go to seed.

Michael Skunda of Burton, Michigan, takes an unorthodox attitude toward weed control. He says his carefully prepared, deeply dug garden bed is so fertile it allows mowed weeds to act as a living mulch that keeps soil moisture even. Skunda waters the plants more often than usual, keeps weeds mowed or clipped to no more than 2 inches high, and says his tomato crop is invariably more bountiful than it was in the days when he weeded his garden.

STAKING AND PRUNING

Gardeners who want to grow tomatoes with as little work as possible can let the plants take their natural form and sprawl on the ground. The vines produce a larger crop, in fact, if not subjected to the pruning that is necessary to train them up a stake. But even the determinate or bush types that do not take kindly to pruning will benefit by being grown next to some kind of support, whether a wire cage, a trellis, or a stake. For one thing, fruits will not rest on the soil, where they risk rot or insect damage. For another, raising tomatoes into the air saves space—and space is usually at a premium in backyard gardens.

Stakes are the usual devices for supporting tomatoes. Staked plants usually have to be trained to no more than four stems, which means they must be regularly pruned (an easy task, since it involves merely pinching out "suckers" when you look over your plants every few days). Indeterminate types benefit from staking and pruning because pruning makes their fruits larger and helps them to maintain their size as the season progresses. The yield of a pruned plant will be smaller, but gardeners can compensate for this by growing an extra plant or two.

Drive stakes 1 foot into the ground about 4 inches from newly transplanted seedlings; putting stakes in place later, when the plant has started its rapid growth, can damage roots. Wooden or bamboo poles, sold at garden stores, are the most commonly used stakes. Some gardeners use metal or fiberglass rods, or even thin, straight tree branches. Anything will do, as long as it is sturdy and tall enough to support the mature fruit-laden vine. Indeterminate tomatoes usually need stakes at least 7 feet long— 1 foot below the ground and 6 feet above.

Tie the plant to the stake every time it grows a foot or so. Use a soft material that will not cut through the stem as the plant moves in the wind: soft baling twine, strips of cloth and even strips of stretchy nylon from pantyhose are suitable. Tie the material in a figure eight—tightly around the stake, but loosely around the stem.

If you are growing an indeterminate variety and choose not to prune it in order to obtain a larger, though smaller fruited, crop, place a stake on either side of the plant and tie the branches to both. If plants are especially vigorous, and no pruning is done, you may need three or even four stakes as supports.

Wire cages provide the most trouble-free method of supporting tomatoes. Plants grown in cages are usually left unpruned, and branches rest on the wire and do not have to be tied. Cages also have the advantage of crowding the leaves together so that fruits are protected from sunscald. See illustration on page 164.

Tomato cages are sold at garden stores, but most gardeners find it easy to make their own cages from concrete reinforcing wire, available at building supply stores. Buy 5-foot-wide wire that has a 6-inch mesh, so that you can reach through to pick tomatoes from the vine. Form the wire into a circle 24 inches in diameter, snipping out the last row of vertical wires so that you can fasten the two ends together. To place the cage around a seedling, snip out the bottom row of horizontal wires and push the cage into the ground. You can make it sturdier by weaving two 3-foot stakes through the mesh and hammering them 1 foot into the soil.

R. D. Waters of Corsicana, Texas, makes use of an otherwise wasted 4½-X-2-foot section of yard between a garage wall and a fence by growing four bush-type tomatoes in a low, rectangular cage. He joins the ends of a 12-foot-X-30-inch piece of reinforcing wire together to form a large circle, then flattens it into a 56-X-16-inch rectangle. Placed over a row of four seedlings, the cage keeps the plants from sprawling into the traveled part of the yard and keeps them out of sight in a garden otherwise devoted to ornamentals.

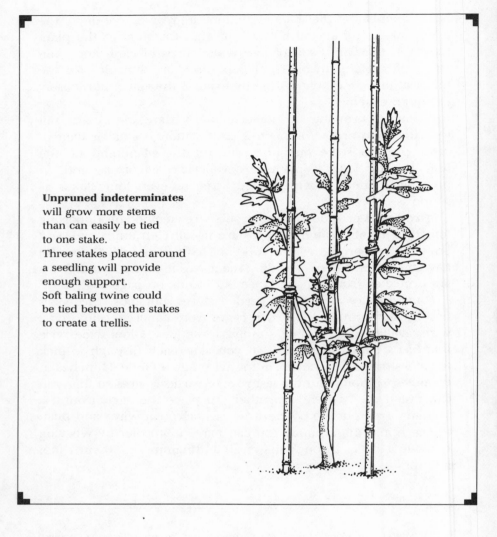

Unpruned indeterminates
will grow more stems
than can easily be tied
to one stake.
Three stakes placed around
a seedling will provide
enough support.
Soft baling twine could
be tied between the stakes
to create a trellis.

Wire cages can also help save space in the garden, especially if the gardener is growing determinates that are not to be staked or pruned.

Trellises can be made from stakes and strings, bamboo poles or lattice. Tomatoes grown on trellises will have to be pruned and trained more meticulously than those grown on stakes, but will produce ripe fruits a few days earlier than plants grown on other supports. See illustration on page 165.

Perhaps the most trouble-free way to trellis tomatoes is to use one of the new nylon trellises, available at nurseries and garden stores. The trellis has a 6-inch mesh and is stretched between two 5-foot posts. Vines are easily woven through the mesh by hand as they grow.

Grover Alexander of Seattle, Washington, constructed a simple bamboo trellis that has been used every summer for six years; in winter it stores flat against a garage wall and takes up little space. (See page 165.)

Alexander tied four 9-foot horizontal bamboo poles to four 4½-foot vertical poles to form a grid, spacing the horizontals at 1-foot intervals from the tops of the verticals. Before Alexander sets out his tomato plants, he drives the vertical poles of the trellis 6 inches into the ground and braces the trellis with two 7-foot poles at each end. As the tomato plants grow, he trains them to three stems each, pinching out all other side shoots and weaving the stems through the horizontal bars. When the plants grow taller than the top bar, Alexander top-prunes them to direct energy into the production of fruits instead of foliage.

Lela Barcroft of Buffalo Grove, Illinois, uses a brightly painted step ladder as a decorative A-frame trellis for the four determinate New Yorker plants she grows every year. She plants one seedling a few inches from each leg, and loosely ties them as they begin to grow. As the vines grow larger, they are supported by the slats and brackets of the ladder, and no fruits are lost by rotting on the soil.

PRUNING

Plants grown on one stake need to be pruned regularly. Doing so trains them to fewer stems and directs energy to the fruits instead of the foliage. There is no mystery to pruning: it is as

163

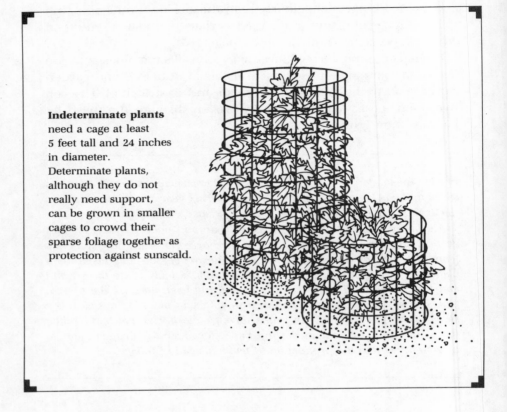

Indeterminate plants need a cage at least 5 feet tall and 24 inches in diameter. Determinate plants, although they do not really need support, can be grown in smaller cages to crowd their sparse foliage together as protection against sunscald.

simple as pinching out "suckers"—the small side shoots that appear in the axils between a main stem and its branches. If the suckers are left to grow, they will form a new branch. See illustration on page 166.

Pinching out enough suckers to train a tomato plant to only one stem will result in an earlier harvest—sometimes as much as two weeks earlier, in fact—but will greatly reduce the number of fruits. Such severe pruning will also subject the fruits to sunscald, cracking, and even blossom end rot. Pruning less, to train the plant to two, three or four stems, will provide better foliage cover for the fruits and increase the eventual yield.

The tops of plants can also be pruned if they outgrow stakes or cages. To keep plants from growing any taller, pinch off the tips of the main stems just above the highest blossom.

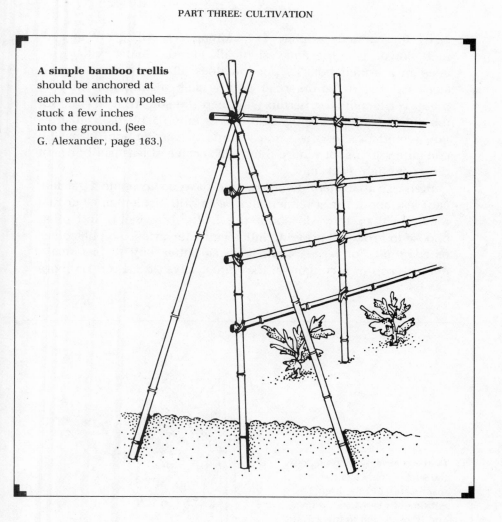

A simple bamboo trellis should be anchored at each end with two poles stuck a few inches into the ground. (See G. Alexander, page 163.)

CONTAINER GROWING

Tomatoes were grown as ornamentals before they were used as food, and modern gardeners still use potted dwarf varieties as decorative accents on patios or sunny porches. Hanging baskets or window boxes planted with miniatures such as Tiny Tim or Basket King are particularly attractive, and larger varieties grown in planters look good when espaliered up the wall of a courtyard or deck.

Five-gallon planters are sufficient for the varieties bred especially for containers (page 114). Miniatures will grow in a 6-inch flower pot; the smallest of all, Florida Petite, will thrive even in a 4-inch pot. To pot tomatoes, fill the container with sterilized soil or commercial potting soil; mixing it with peat moss, vermiculite, or perlite will keep the soil from drying out too quickly after watering. Add half as much 5-10-10 fertilizer as you would use for the same volume of garden soil. Place the containers in a spot where they will receive at least six hours of direct sunlight each day.

High-rise apartment dwellers, who have no access to a garden, care less about the ornamental aspects of their container-grown tomatoes than they do about the fruits they yield, and often choose to grow tall-growing indeterminate varieties on balconies or rooftops. For large plants such as Better Boy to produce a decent crop in a container, they must have 20 gallons or more

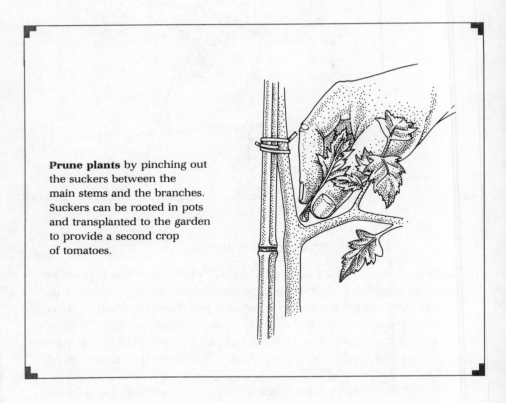

Prune plants by pinching out the suckers between the main stems and the branches. Suckers can be rooted in pots and transplanted to the garden to provide a second crop of tomatoes.

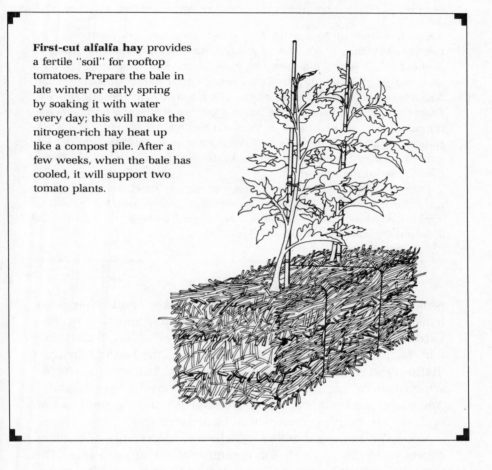

First-cut alfalfa hay provides a fertile "soil" for rooftop tomatoes. Prepare the bale in late winter or early spring by soaking it with water every day; this will make the nitrogen-rich hay heat up like a compost pile. After a few weeks, when the bale has cooled, it will support two tomato plants.

of soil; and as they grow, they need to be supported by stakes or a wire cage.

PROBLEMS, PESTS, AND DISEASES

Tomatoes have a reputation for being easy to grow. But gardeners who hear of all the problems tomatoes are heir to—from bugs to fungus diseases to scalding by hot sun—may wonder how anyone gets a crop at all. Those who get discouraged should

Tom McNamee of New York City grows tomatoes on the roof of his Greenwich Village apartment in a novel way: his container is a bale of first-cut alfalfa hay, a nitrogen-rich growing medium that heats up like a compost pile. Beginning the first week in March, McNamee gives the bale a good daily soaking with water, which activates the heating process. Over the next 7 or 8 weeks, the bale decays to a fertile compost, although it maintains its shape; it also cools down enough for planting. McNamee stands two stakes in the bale and nestles a tomato seedling into the hay next to each stake. A daily watering keeps the plants growing vigorously for the rest of the season.

Other types of hay, such as fescue, can also be used, but will not heat up to form a nutrient-rich growing medium. Unlike tomatoes grown in alfalfa hay, those grown in fescue or other hays will have to be fed almost daily with a weak soluble fertilizer.

remember that, despite myriad pests, diseases, and other problems, things are not as bad as they seem: the tomatoes in their yards are not about to be attacked from all sides. Verticillium wilt, for example, the fungus disease that is the bane of growers in the West, is almost unknown in the South. Tomato fruitworms, which can ravage plants in Georgia, are rare in New England. And many physiological problems, such as cracking and blossom end rot, can be prevented without much trouble.

Moreover, home gardeners have benefited from plant breeders' efforts to improve yields for commercial tomato growers. The verticillium, fusarium, and nematode (VFN) resistance so necessary for the success of commercial crops has also been bred into a number of home garden varieties; and dozens of resistance-packed commercial varieties, which counter physiological disorders as well as diseases, are found by many gardeners to be flavorful enough for growing at home.

PHYSIOLOGICAL DISORDERS

Some tomato problems are not caused by organisms or diseases, but are the consequences of adverse weather. Almost all of these physiological disorders affect fruits rather than foliage.

Blossom drop is perhaps the most frustrating. The tomato plant's small yellow blossoms break off and drop to the ground before they can set fruit, usually because the temperature is too high or too low. Most midseason varieties will not set fruit when night temperatures fall below 55° or rise above 75°. (Early varieties have the advantage of setting fruit at lower night temperatures, and many hot weather varieties have been developed that will continue to set fruit on hot nights in the South.) These temperature extremes inhibit fertilization of the blossoms; and if fertilization does not take place within two or three days of the blooming of the flower, the blossom drops off. The cost to the gardener is one tomato per fallen blossom—a high price to pay.

Even when night temperatures are within the acceptable range of 55° to 75°, a steady wind that is hot and dry can cause blossoms to drop. Or heavy rains can knock the fragile flowers to the ground. Even too-dry soil or too much nitrogen fertilizer can aggravate the problem.

It is easier to control blossom drop in cool climates than in hot ones, since temporary covers—hotkaps, plastic milk jugs, or arched fiberglass sheeting—can act as minigreenhouses to protect blooming young plants on cold nights (page 154).

Some gardeners use a commercially available spray that chemically fertilizes the blossoms before they can drop. More often than not, the sprayed blossoms give birth to hard, seedless, vaguely tomatolike monstrosities that are hardly what the unsuspecting gardener had in mind when he set out to grow tomatoes.

Blossom end rot looks like a disease, but is actually caused by the tomato plant's temporary inability to take up calcium. A dark, leathery spot, as small as a dime or as large as half the fruit, appears at the blossom end of the tomato. The calcium deficiency that causes it can be the result of uneven moisture, excessive nitrogen, or too-high acidity of the soil. The disorder is most common in plants that have been staked and pruned.

Uneven moisture—a sudden dry spell after a run of wet weather, or vice versa—can interfere with the way the plant absorbs calcium from the soil. Careful watering, mulching, and adequate drainage will usually prevent the problem. Using a fertilizer high in nitrogen, such as blood meal, cottonseed meal or any chemical fertilizer with as much nitrogen as phosphorus and potassium (a 10-10-10, for example), can also interfere with calcium absorption, as can highly acidic soil. Acidity can easily be measured by a soil test (page 130) and corrected by raking lime into the soil before setting out tomato plants.

Tom Priest of Downingtown, Pennsylvania, uses epsom salts to help prevent blossom end rot in his tomatoes. He simply throws a handful into the planting hole and covers it with a thin layer of soil before putting in the plant. (Epsom salts, actually magnesium sulfate, have been shown in scientific tests to aid the transport of calcium to the tops of tomato plants, and thus to the fruits. Besides helping to overcome the calcium deficiency that can cause blossom end rot, the salts also help the plants absorb phosphorus and sulphur.)

Catfacing is a malformation of the tomato. It is caused at an early stage, when part of the blossom sticks to the tiny, developing tomato and disfigures it as it grows. Cloudy, cool days contribute to the problem. Despite the condition's graphic name, the affected tomato looks nothing like a cat's face (nor anything like a tomato,

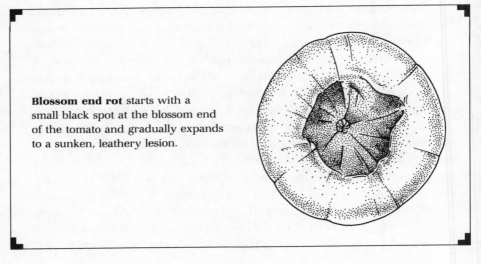

Blossom end rot starts with a small black spot at the blossom end of the tomato and gradually expands to a sunken, leathery lesion.

either): the blossom end of the fruit shows puckers, scar tissue, and deep crevices.

Catfacing is rarely a major problem for home gardeners, and usually occurs only in the first fruit set. But those few who are plagued by it may want to grow one of the varieties known to resist it.

Cracking of tomato fruits on the vine can happen in two different ways, and for two different reasons. Concentric cracking, in which the cracks form small rings encircling the stem end of the fruit, is a problem in high rainfall areas. Droplets of water that remain on the fruit for six hours or more can cause the skin of the fruit to split. A second kind, radial cracking, shows deep cracks radiating down from the stem end of the fruit. It is caused by uneven soil moisture that makes the growth rate of the fruit fluctuate. A mulch will help keep the moisture supply even.

Sunscald occurs when sparsely foliaged plants expose their green fruits to hot sun for several days. The sun burns a yellow patch into the side of the fruit. The patch may later turn into a blister, which in turn changes color to grayish-white and becomes dry and papery. The blister itself does not affect the rest of the tomato—the unblemished part of the fruit can be eaten—but it provides a convenient doorway for invading fungus spores.

Sunscald is most common in the Southwest, but can occur anywhere during a spell of hot, dry weather, especially when tomato plants have been defoliated by diseases or pests. Prevention involves maintaining the plant's protective leaf cover. This means, obviously, guarding against foliage diseases; but leaf cover can also be increased by growing a plant in a cage, which crowds its foliage and provides more shade. And, in the Southwest, gardeners may want to grow one of the varieties that have been bred to have heavy foliage (page 115).

Puffiness makes a slice of tomato look like that of a bell pepper—hollow between the outer walls and the seedy center. It is the result of adverse conditions during pollination: extreme temperatures, heavy rain on cold days, or heavy cloud cover that makes for low light. It can also be caused by using a blossom-setting hormone spray. Puffiness is most likely to occur in tomatoes harvested early in the season.

Nippling gives tomatoes a pointed blossom end, so that they are shaped like a plumber's bob. The abnormality is usually the result of cooler-than-average nights when the tomato blossoms are being fertilized. The problem is merely cosmetic; it does not affect the taste or quality of the fruits.

Leaf roll, in which leaves of the tomato plant curl up and feel leathery to the touch, is often mistaken for a disease. While some foliage diseases cause the leaves of tomato plants to curl, leaf roll without yellowing, spots, or leaf death is simply caused by heavy pruning, or by heavy rains that leave the soil too wet for a few days. It does not affect fruit production in any way. Gardeners who are concerned about it from a cosmetic standpoint may find they can alleviate the problem by pruning less severely, or by mulching plants to keep soil moisture uniform.

PESTS

If the control of voracious bugs and worms and flying insects in the garden is not any easier than it used to be, it is at least safer. The hazardous, highly residual sprays and powders of the past have been outlawed, and the comparatively innocuous chemicals that remain can be coupled with a number of organic, or nonchemical, methods to keep pests away from the tomato patch.

Gardeners in the South, where pests can survive and multiply through mild winters, usually have to take stronger measures than do northern gardeners. Without some kind of chemical control, they can lose a huge portion of their tomato crops. And some organic methods simply will not work in hot climates. Companion planting, for example—interplanting tomatoes with supposedly pest-repelling plants such as basil or catnip—have been shown in tests by entomologists in Georgia to be completely ineffective (an exception is the case of nematode-controlling marigolds; see page 182). Basil next to tomatoes, in fact, was found to actually attract hornworms and cabbage loopers.

The five pesticides recommended on the following pages are among the most benign. Four of them—Sevin, Malathion, Diazinon, and Resmethrin—are decidedly mild, and remain in the soil for no longer than ten days. The fifth is not a chemical at all, but a biological insecticide containing the natural bacterium *Bacillus thuringiensis.* Sometimes called BT, it is sold as Dipel. Sprayed on tomato foliage or soaked into the soil around plants, it is ingested by—and eventually kills—tomato fruitworms, horn-worms, and cutworms.

Cutworms may be gray, brown, black, striped, or spotted, according to their species. But their colors and markings are of no importance to the gardener, since he never sees them. The sly creatures hide under clods and mulch in the light of day, and emerge only at night to do their nasty work—toppling young tomato plants by cutting cleanly through their stems.

The most widely practiced method for controlling cutworms is to place some kind of collar—a tin can or plastic cup with the bottom cut out—around young plants. Most species of cutworms cannot surmount the barrier. This method protects the plants from worms outside the collar, but not from any that may be lurking within the few square inches of soil *within* the collar, or from soil-borne eggs next to the stem. A much more effective control is to wrap the stems of seedlings, before setting them out in the garden, with a 4-×-4-inch piece of aluminum foil. The seedlings then are planted with 2 inches of foil above the soil surface and 2 inches below. (The foil should be loose enough to allow the stem to expand as the plant grows.) This method also protects plants from the soil-borne fungus spores of southern blight.

A few species of cutworms can climb, and are not discouraged by the foil collar. If these worms show up in the garden, they can be controlled by soaking the soil around the base of the plants with Dipel or Sevin.

Tomato fruitworms are the most common pest in the deep South, and are troublesome for gardeners westward through the Southwest to California. The same caterpillar is known by a number of different names, depending on the plants it is attacking: corn-ear worm, soybean-pod worm, and cotton-boll worm are three. In the tomato patch, the caterpillar burrows into green or ripe fruits. The gardener is lucky if the pest confines its meals to one or two tomatoes, but often the restless pest will crawl from fruit to fruit, burrowing briefly into each and causing the fruit to rot.

The caterpillars are yellow to brown in color and measure up to 1½ inches. A spray of Sevin at the plants' first bloom, followed by reapplications every 7 to 10 days, will control them. Dipel, too, is effective, but because it takes longer to work, the gardener

Tomato fruitworms are yellow or dark brown and have alternating light and dark stripes running lengthwise along the body.

should apply it earlier, while the caterpillars are still small. It should be reapplied every 7 days.

A similar pest, the fall army worm, causes the same kind of damage to tomatoes. It can be distinguished from the tomato fruitworm by its black head with an inverted white Y on the front. Unlike the fruitworm, it feeds at night. Sevin, applied every 7 days, will control it; Dipel, because it is slower-acting, is not as effective.

Hornworms are large caterpillars—3 to 4 inches long—with an intimidating horn that, surprisingly, is at the worm's tail end instead of the head. The hungry worms can eat so many leaves that they stunt plants and expose green tomatoes to sunscald.

Hornworms are less threatening than they look: they do not sting, and the simplest control is to pick them off plants by hand. Some gardeners, to avoid touching them, snip them in half with scissors. Another control is the application of Dipel, which takes a few days to work.

Wasp pupae often feed on hornworms. Once the worm is dead, the parasites on its body are often mistaken for hornworm eggs by gardeners, who then destroy the parasites and thus do away with friendly insects that keep the hornworm population under control naturally.

Cabbage loopers are smooth-skinned green worms, up to 1¼ inches long, that move along like inchworms. They chew leaves and can defoliate tomato plants. Dipel, used weekly, will control them if the spray program is started early enough in the season.

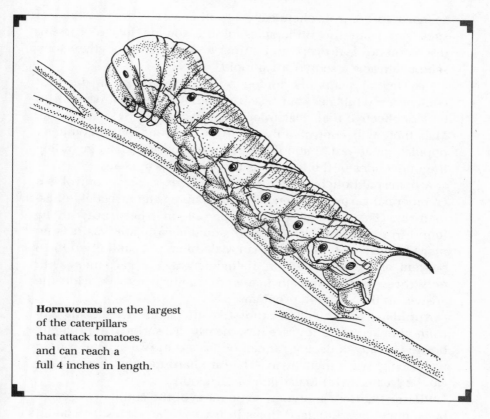

Hornworms are the largest
of the caterpillars
that attack tomatoes,
and can reach a
full 4 inches in length.

Because hornworms are so well camouflaged on tomato plants, they are usually hard to spot. Terry Schoen of Walla Walla, Washington, discovered an easy way to locate the worms on a hot summer day: he simply directs a fine spray of water from the hose at his plants. This cold spray causes the worms to thrash about, shaking the leaves around them. Once spotted, they can easily be picked off.

Whiteflies are a frustrating problem for most tomato gardeners—partly because they attack plants in such numbers (a plant on which they are feeding will be surrounded by a thick swarm of whiteflies when shaken), and partly because they are

175

so hard to control, chemically or otherwise. The tiny insects suck sap from the undersides of the tomato leaves, causing discoloration, leaf drop, and stunting. And the honeydew they secrete attracts a sooty fungal mold.

Insecticidal soaps are touted by manufacturers as the best control for whiteflies, but tests by entomologists show them to be less effective than Malathion, especially in hot climates. Still, Malathion will control only 40 to 50 percent of the whitefly population, at best. It should be sprayed on the plants, including the undersides of the leaves, as the whiteflies appear.

A better, although almost prohibitively expensive, control is a commercial aerosol garden spray containing the insecticide Resmethrin. (The name is rarely apparent on labels, but can be found in the list of ingredients of some aerosol sprays.) It is no more hazardous to humans than Malathion, yet can kill up to 70 percent of the whiteflies on a single plant. For gardeners with only three or four tomato plants, Resmethrin may provide the answer to the whitefly problem.

Aphids, like whiteflies, stunt tomato plants by sucking sap from the leaves. They are not nearly as serious a problem, however, since a decent rainfall or stream of water from a hose will usually wash them away. The rare hard-to-control infestation can be treated with Malathion or Diazinon.

Blister beetles, known in some parts of the country as potato bugs, chew large, ragged holes in leaves. They can defoliate an entire tomato plant, or at least expose the fruits to sunscald. The beetles, most common in arid regions, are ¾ to 1 inch long and come in various colors. They take their name from the fact that they cause severe irritation, even blistering, to the skin when crushed. They can be controlled by spraying or dusting with Sevin.

Colorado potato beetles chew on tomato leaves in both the adult and larval stages. As adults, the hard-shelled beetles are about 1½ inches long, and are red with black stripes; as larvae, they are red, humpbacked grubs marked with rows of black spots. Both the adults and the worms can defoliate tomato seedlings early in the season.

In the northern states, where Colorado potato beetles are most common, the pests have developed resistance to Sevin, the traditional treatment. Gardeners there should pick the beetles off the plants by hand, being careful to dispose of any yellow egg masses as well. In the South, a spray of Sevin, followed by a second spray a few days later, will control the beetles.

A homemade spray made from cedar boughs will kill Colorado potato beetles in their larval stage. Lesley R. Hickerson of Durango, Colorado, learned the method from his mother in Oklahoma. She put pieces of cedar boughs, both foliage and wood, in a large kettle and boiled them until the water turned brown and oily. The liquid was then sprinkled on infested potato and tomato plants, where it killed the worms on contact. Nowadays, gardeners can use a spray gun for better coverage.

Flea beetles attack the leaves of tomato seedlings until the first fruits set. The tiny black beetles, no more than ⅛ inch long, jump like fleas when disturbed. They descend on the garden in groups, and are attracted by weeds in or nearby the garden. The first line of defense is to keep the garden and surrounding area weed-free. Infestations can also be prevented by one application of Sevin around the edges of the garden, to form a barrier, when seedlings are set out.

Stinkbugs worry gardeners more than they should, since their damage to tomatoes is only cosmetic. They withdraw juice in minute amounts from green tomatoes, and when the tomatoes ripen, the tiny spots where the bug penetrated turn yellow. The spots do not make the tomato any less edible.

The adult bugs are bright green or brown, flat, and shield-shaped, measuring about ⅝ inch long. They smell bad when crushed. Younger bugs are more rounded, are wingless, and are green, orange, or black in color. Stinkbugs are a late-season pest, and are attracted by weeds. If keeping the garden weed-free does not control them, spray with Sevin every 7 days.

Leafminers are likely to be a problem only in those gardens where harsh, broad spectrum pesticides have killed off the parasites that keep leafminers in check. The tiny maggots make winding white trails through tomato leaves, making the foliage susceptible to disease. Diazinon will control them.

177

Two-spotted spider mites, also called red spider mites, can be troublesome in arid regions. The barely visible mites suck juice from the undersides of tomato leaves and weave fine webs on the foliage. High-pressure spraying with water reduces the population, but serious infestations are better controlled with Malathion.

Squirrels are among the most frustrating pests, because there is no truly reliable way to repel them short of standing in the middle of the garden twenty-four hours a day. In some years, they can pick or bite virtually every tomato in a garden, wiping out an entire crop with more efficiency than any bug or disease. Gardeners who have suffered that misfortune can take heart: squirrels do not attack tomatoes every year as a matter of course; they usually develop a taste for tomatoes only when their routine foodstuffs are in short supply for one reason or another. So the gardener whose tomato patch has been devastated one year should not hesitate to try again the next.

Methods that have been tried to discourage squirrels include hanging tin cans in the garden to clatter in the wind, sprinkling black pepper on the soil, placing rubber snakes in clear view, spreading dog hair around the garden, and keeping an active cat or dog nearby. These methods may work part of the time, but have never been shown, in the long run, to have an effect on the number of tomatoes lost to squirrels. And the squirrel-proof net has yet to be invented.

DISEASES

Gardeners who blame plant breeders for taking the taste out of tomatoes should congratulate them for putting disease resistance into the plants. Most new tomatoes have been bred to ward off the fungal, bacterial, and viral diseases that are almost impossible to control once they start. Some of the varieties may lack the succulence and flavor of older, nonresistant ones, since they are designed primarily to suit the needs of large-scale commercial growers for whom yield, uniformity, firmness, and concentrated fruit set take precedence; but, fortunately, a few new resistant varieties are intended only for the home garden, where the qualities that count are taste and texture.

Seed catalogues, seed packets, and nursery labels indicate the diseases to which tomatoes are resistant by listing the initial letters of those diseases after the variety's name. "Better Boy VFN," for example, shows that Better Boy is resistant to the three most common problems that strike tomatoes. "V" is for verticillium wilt, a fungus disease prevalent in the West and in some of the northern states. "F" is for another fungus disease, fusarium wilt. (There are several species, also known as races, of fusarium, and a single "F" indicates resistance to only the most common race. Some tomatoes resist the second most common race, as well, and their names are followed by "F_1F_2," for fusarium race 1 and fusarium race 2.) "N" is for nematodes—not actually a disease, but tiny underground eelworms that attack the roots of plants. (Nematode damage is classed with the diseases because its symptoms appear to the gardener to have been caused by disease rather than pests.) Some tomato variety names also are followed by the letter "T," which indicates resistance to tobacco mosaic virus. A tomato with any of these four letters listed after its name will stand up to attack by the disease or diseases indicated, but will not necessarily be completely immune.

Some tomatoes are also said to have "tolerance" for certain diseases. A tomato described in a seed catalogue as "tolerant to verticillium wilt," for example, will withstand it better than a tomato with no resistance at all; but it will not have the stronger degree of resistance that earns a "V" after its name. Many tomatoes have been bred to have tolerance for less common diseases, such as gray leaf spot and early blight.

Choosing resistant varieties is obviously the best step the gardener can take to keep the four most common problems— verticillium, fusarium, nematodes, and tobacco mosaic virus— out of the garden. But since resistance is not, by any means, the only consideration for choosing varieties, the gardener often has to fight disease with chemical control or organic methods. As with pesticides, the chemicals recommended on the following pages are among the least hazardous available. They will prevent or control disease in some cases; but in other cases, the gardener may have no choice but to give up on his tomatoes and look to next year.

Verticillium wilt is most common in the West, but occurs in most other parts of the country except the South. It is less serious than fusarium wilt. Plants affected by it will still produce tomatoes, but they will be small; and because the disease defoliates all but the leaves at the ends of the branches, many of the fruits will risk blistering by sunscald.

Verticillium can be diagnosed through a few obvious signs. First, the plants droop. The tips of the growing shoots look slightly wilted during the warmest part of the day, and the older leaves turn yellow, eventually withering and falling. Over days or weeks, the plant becomes defoliated except for the leaves at the ends of the branches. The crown loses all its leaves, and the leaflets remaining on the plant curl up. The plant stays alive; however, a straggly specimen bearing stunted tomatoes that may or may not be worth harvesting.

Verticillium wilt thrives when the temperature is 70–75°, and is retarded by higher temperatures; this is why it is rarely a problem in the searing summers of the South and Southwest. The West, however, is an ideal breeding ground for the disease, and western gardeners will find that related vegetables, including eggplants, peppers, and potatoes, are just as vulnerable as tomatoes.

The main precaution against verticillium wilt is, obviously, to grow one of the many new varieties with inbred resistance to the disease. These usually have a "V" alongside their names in seed catalogues or on nursery labels.

Gardeners who grow nonresistant varieties (far more numerous than the resistant ones, since they include all tomatoes introduced before the late 1950s) can take a few safety measures to guard against verticillium wilt. Because the disease is soil-borne, the soil used for germinating seeds and for transplanting seedlings into larger pots should be sterilized. This is easy to do: simply sprinkle a layer of soil over a foil-covered cookie sheet and place it in a preheated oven at 350° for at least 30 minutes.

Gardeners who have had serious trouble with soil-borne fungus diseases in the past should rotate their crops each year, never growing any plants that are members of the Solanaceae family— eggplants, peppers, okra, and potatoes, as well as tomatoes—in the same place, nor interchanging them, for three growing seasons. (Spores from previous infestations can live in the soil for up to three years.) This may present a problem to gardeners who like to grow three or four, or even all, of these susceptible vegetables, yet have little space. If so, they may have to enlarge their plots by two or three rows, being careful not to mix one part of the garden's soil with another. Or, if the fungus diseases have been really severe, a new garden plot may have to be dug in another part of the yard.

Fusarium wilt is more critical: if it hits tomato plants in their early stage of growth, it will kill them. It is mainly a problem in the South and Southwest, since it thrives when the temperature of the soil and air is 80–90°, but it can appear in all areas in the case of an extended heat wave. Unlike verticillium, it affects only tomatoes, sparing the other members of the *Solanaceae* family.

Fusarium wilt may be hard to tell from verticillium, since it, too, causes wilting and yellowing of leaves. The gardener can look for a couple of distinguishing characteristics, however: verticillium, in its early stages, affects the whole plant; fusarium may at first affect only one shoot, or the leaves on one side of a shoot. And, instead of leaving the foliage at the ends of the growing stems intact, fusarium will gradually ravage the entire plant, progressing up the stem until all the leaves are killed and the stem dies. A sure way to identify fusarium wilt is to cut the main stem of an affected plant lengthwise; the woody part of the interior will show dark discoloration.

Diagnosis—finding out whether wilting is caused by fusarium or verticillium—may seem useless to the gardener, since he knows he can do nothing to remedy it once it starts. But it at least tells him which inbred resistance to look for when choosing varieties for the next year.

The most common form of fusarium wilt is known as "race 1" to distinguish it from the several other species of the disease. In recent years, scientists have bred resistance to the second most common species, known as "race 2," into some tomatoes. Varieties with resistance to both races are designated "F_1F_2."

Making sure tomato plants do not suffer any root damage may go a small way toward keeping fusarium from taking hold in the garden. Cultivating the soil around plants with a spade is not advisable—it can easily injure roots and provide a portal for entry of the disease.

Nematodes are tiny, almost microscopic eelworms, sometimes called root knot eelworms. Several species of the pests are found throughout the U.S. and Canada, and they are so well adapted to their regional environments that they are likely to be a

181

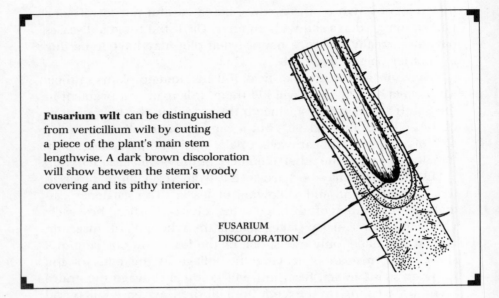

Fusarium wilt can be distinguished
from verticillium wilt by cutting
a piece of the plant's main stem
lengthwise. A dark brown discoloration
will show between the stem's woody
covering and its pithy interior.

FUSARIUM
DISCOLORATION

problem in any backyard on the continent. They infest soil and
mount a full attack on the roots of vegetables. Not only tomatoes
are affected, but also cucumbers, melons, squash, eggplants,
lettuce, spinach, carrots, and celery. Affected plants weaken and
eventually die.

The aboveground symptoms of nematode infestation in tomatoes are stunting of the plant, despite adequate water and
fertilizer, and wilting—without yellowing of leaves—during the
hottest part of the day. The one unmistakable symptom, however,
can be seen at the site of the infestation itself: the roots. They
show elongated, round swellings, called root knots or galls.

Many newer varieties of tomatoes are resistant to nematodes.
Their varietal names are followed by the letter "N" in catalogues
and on nursery labels. Unlike the other two common tomato
problems, verticillium and fusarium wilts, nematodes can be
controlled.

The solution is to start a prevention program a year ahead,
planting marigolds as a cover crop in the part of the garden
where tomatoes are to be grown the next year.

Most organic gardeners are aware that marigolds control

nematodes, but have the idea that simply interplanting them with tomatoes will do the trick. Actually, such interplanting has no effect at all. Even planting tomatoes and marigolds in the same hole will do nothing to discourage the pests—they will simply ignore the marigolds and attack the tomato roots.

When a portion of the garden is given over entirely to marigolds, however, any nematodes in the soil have no choice but to feed on their roots. A toxin in the roots kills them, and the soil is left nematode-free for next year's tomato plants. (The gardener need not worry about nematodes invading the purged soil from another part of the garden—nematodes live their entire lives within one inch of where they were hatched.)

When planting marigolds as a cover crop, place them no more than 6 inches apart, and immediately remove any weeds that spring up between them during the growing season. Left to grow for an entire summer, the colorful flowers will rid the soil of nematodes as effectively as harsh chemical nematocides.

Tobacco mosaic virus attacks not only tobacco, but also plants that are related to it, including tomatoes, eggplants, peppers, and potatoes. Symptoms on tomato plants are light green or dark green mottling of the leaves, with curling and malformation of the leaflets. Some green fruits may have brown spots.

Unless the virus attacks young seedlings, it is not as harmful as verticillium, fusarium, or nematodes. But it affects the vigor of the plant and will reduce fruit yields by about 20 percent; and, some fruits may be lost because of spotting.

The virus is extremely infectious, and is easily spread from one plant to another by the gardener's hands. It is so persistent that it exists in the processed tobacco of cigarettes and cigars, and is transferred to the hands of smokers when they touch them. Washing the hands with plain soap and water will not control the virus, but washing them well with laundry detergent will. Strangely enough, sweet milk is also good for inhibiting the virus. When handling infected plants, gardeners can help prevent spreading the disease by dipping their hands in milk before touching other plants.

A few newer varieties of tomatoes, such as Celebrity, Big Pick, and Park's Extra Early, have inbred resistance to tobacco mosaic virus. Resistance is indicated by the letter "T" listed after the variety's name.

Early blight is a common disease in humid areas, affecting both the foliage and fruits of tomato plants. It is also known as alternaria stem canker, after the fungus that causes it, *Alternaria*. Although the disease occurs in a wide range of climates, it is encouraged by heavy dews and rainfall. It is not common in arid parts of the West.

Early blight, despite its name, usually affects older plants. (It is found in seedlings, too, but not as often.) Dark brown spots with concentric rings develop on the older leaves. Spotted leaves may wither and die and partially defoliate the plant, resulting in sunscald of the fruits. Worse, the fruits will develop a black, leathery sunken area at the stem end, often with dark concentric rings.

The fungus can live in the residue of diseased plants over the winter, surviving for at least a year. Plants that are lost to the disease should be burned.

Early blight can be controlled with applications of the fungicides Maneb, Zineb, or Mancozeb. Applications should be repeated after rains.

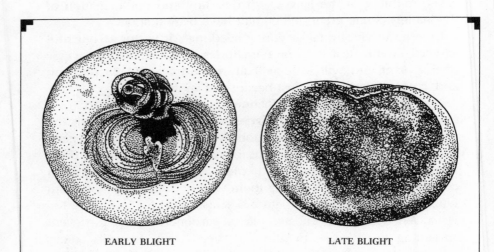

EARLY BLIGHT   LATE BLIGHT

Blights cause tomatoes to rot. Early blight starts at the stem end and gradually enlarges in concentric rings. Late blight shows dark brown, wrinkled spots over most of the fruit.

Late blight, like early blight, occurs in humid areas. It is caused by a different fungus, *Phytophthora,* and is less common. It affects potatoes more often than tomatoes.

Symptoms appear on both foliage and fruits. Irregular, greasy, gray areas develop on the leaves, enlarging quickly during wet weather. A white downy mold appears at the margin of the affected area on the lower surface of the leaves. (This downy mold distinguishes late blight from early blight; diagnosing the blight is not particularly important for the home gardener, however, since the treatment for both diseases is the same.) Gray-green, water-soaked spots appear on green tomatoes and gradually grow larger in irregular shapes. The spots become dark brown, firm, and wrinkled.

The disease-causing fungus will last through winter in frost-free areas of the South, and will survive in northern climates in potato cull piles, ready to reinfect the garden in spring. Like early blight, its spread is promoted by wet weather and cool temperatures. It does not occur in hot, dry climates.

Late blight can be controlled with applications of Maneb, Zineb, or Mancozeb. Applications should be repeated after a rainfall.

Southern blight is another soil-borne fungus disease. The first symptom is wilting of the plants. The stem at the soil line shows a brown, soft rot, usually covered with a white, cottony mold embedded with tiny brown spots. Eventually the plant dies, often without any yellowing of the leaves. If fruits touch the soil, they develop yellow, sunken areas that soon break open.

Since the fungus grows very little at temperatures below 68°, it rarely appears in cold regions. It is most common in the South, in gardens with light, poorly drained soil. The disease can be prevented with the same aluminum foil collar that prevents cutworm damage (page 173).

Anthracnose differs from the other blights in that it affects tomatoes only when they are ripe. It is most common east of the Mississippi. Fruits show sunken, water-soaked spots, measuring up to ½ inch in diameter; they are sometimes surrounded by concentric rings. The tan centers of the spots may eventually erupt with a mass of salmon-colored fungal spores. Older leaves on the plant may be affected, showing small brown and yellow spots.

The fungus that causes anthracnose, *Colletotrichum*, can last through the winter on plant debris, and thrives in poorly drained soils. Crops should be rotated so that tomatoes are not planted in infected soil for 3 to 4 years.

To control the disease, spray or dust with Maneb, Zineb, or Mancozeb, repeating the applications after a rainfall.

Gray leaf spot, a common southern disease sometimes called *Stemphylium* leaf spot, after the fungus that causes it, attacks only the leaves of the tomato plant. Older leaves are affected first. Several small, dark brown spots appear and extend through to the undersurface of the leaf. The spots gradually enlarge and take on a gray-brown, glazed appearance. The centers of the spots often crack, and eventually fall out entirely, creating holes in the leaves. When spots cover most of the leaf, the leaf will turn yellow, wither, and fall. Severely affected plants will become defoliated except near the ends of the growing shoots, and few fruits will be produced.

Debris from diseased plants harbors the fungus, which can remain in the soil for several years. For this reason, diseased plants should be burned. Tomatoes and related plants, such as peppers and eggplants, should be rotated to another part of the garden for the next two or three growing seasons.

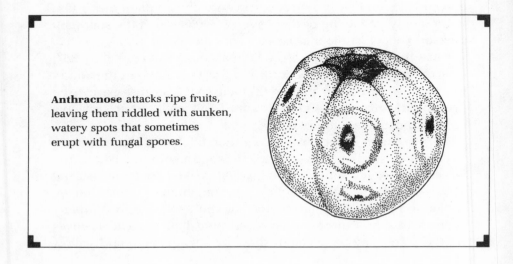

Anthracnose attacks ripe fruits, leaving them riddled with sunken, watery spots that sometimes erupt with fungal spores.

Applications of Mancozeb will help control gray leaf spot. Some varieties of tomatoes, mostly southern-bred, have some tolerance to the disease. Among the more commonly grown are Walter, Manapal, Floradel, Manalucie, and a greenhouse variety, Tropic.

Septoria leaf spot, caused by the *Septoria* fungus, is characterized by several small gray, round leaf spots with dark borders. A few black, pinhead-sized dots may be seen within the spots; these are *pycnidia*, the reproductive structures of the fungus. Like gray leaf spot, the disease survives in residues from diseased plants. Wet weather favors its growth and spread, which can result in defoliation of the plant.

The disease can thrive anywhere when temperatures are moderate and rainfall is heavy. Mancozeb will control it.

Damping off is a common problem with tomato seedlings in the first two weeks after they emerge from the soil, either in the garden bed or in pots. A fungus attacks the seedling at the soil line. Usually, the roots are killed and the affected plants show water-soaking and shriveling of the stems at ground level; they soon fall over.

Excess moisture allows the fungus to take hold. Seedlings are susceptible for about two weeks after they germinate. As their stems harden and increase in size, they become less likely to succumb.

To prevent damping off, make sure the drainage of germination pots, or of garden soil, is adequate. Sterilize the soil for germination in an oven at 350° for 30 minutes. Do not overwater the seedlings, and make sure they have good ventilation. If sowing seed directly in the garden, dust the soil with a fungicide such as Maneb or Zineb.

Bacterial wilt can easily be distinguished from the more common verticillium and fusarium wilts, since affected tomato plants will wilt suddenly without yellowing. The pith in the centers of the stems becomes water-soaked and later turns brown. Within days the stem will have become hollow and the plant will die.

The disease is caused by bacteria called *Pseudomonas solanacearum,* which are borne in the soil. It is most prevalent in the South, in gardens where the soil is constantly moist and the temperature is usually above 75°. It attacks not only tomatoes, but also potatoes, peppers, tobacco, peanuts, and soybeans.

Like other wilts, it cannot be effectively treated. Diseased plants should be uprooted and burned. Insect control also helps retard the spread. Crop rotation should be practiced, without planting tomatoes, peppers, eggplants, or potatoes in the affected soil for several years. The best solution is to move the entire garden to another part of the yard.

Bacterial canker is borne in tomato seeds. The surest method of control is to buy treated seeds that are guaranteed to be free of disease. However, the bacteria can live in the soil, and will attack tomatoes grown in infected areas.

Bacterial canker is characterized by wilting leaflets on old and young plants alike. On older plants, the leaves die from the margin inward, toward the midrib. Often, only one side of the leaf will be affected at first; eventually the whole leaf will shrivel and die. Many plants will survive the disease, but they will be stunted, and will produce few tomatoes. Fruits that are attacked by the bacteria will at first have small, white spots that soon develop dark centers surrounded by a white halo.

There is no effective control. Even if the gardener moves his tomato patch the next year, he should not use the same stakes that supported infected plants, since they can carry the disease.

Bacterial spot damages both foliage and fruits in states east of the Mississippi, especially where rainfall is high. It is most apparent on green fruits, which develop water-soaked spots that measure up to ¼ inch in diameter. Like the spots caused by bacterial canker, they are surrounded by a white halo—but they are raised above the surface, while bacterial canker spots are relatively flat. On foliage, leaflets have dark, greasy spots; when older leaves are affected, there may be defoliation. Bacterial spot can be controlled with applications of a compound that contains Maneb, zinc, and copper.

Bacterial speck is similar to bacterial spot, but affects only young fruits. The raised spots it creates on the fruits measure less than ¹⁄₁₆ inch in diameter. Leaves of infected plants develop similar spots. Infection is most likely to occur when heavy rains splash the bacteria to all parts of the plants. Control is the same as for bacterial spot.

Curly top virus occurs in the Pacific Northwest, and is carried

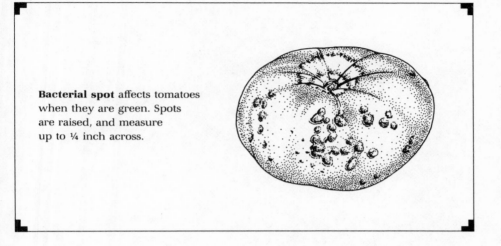

Bacterial spot affects tomatoes when they are green. Spots are raised, and measure up to ¼ inch across.

from plant to plant by an insect called the sugar beet leafhopper. The disease causes tomato plant foliage to curl up and parts of the roots to die, resulting in severe stunting or death of the plant. The insect that carries the disease can be controlled with Sevin. Or, the gardener may want to plant one of the varieties bred to resist the disease, such as Roza.

Sclerotinia stem rot is caused by a fungus that also attacks beans, cabbage, carrots, celery, cucumbers, lettuce, onions, peas, pumpkins, and squash. A dry rot encircles the stem at ground level and eventually kills the plant.

The disease usually occurs west of the Mississippi, and thrives during long cool spells. There is no effective control, but gardeners can help prevent the disease by making sure the garden bed is well-drained.

Walnut wilt affects tomatoes when they are grown in the root zone of trees of the *Juglans* genus, commonly known as walnut and butternut trees. A toxin, called juglone, leaches from the roots of the tree and causes the tomato's woody stems to turn brown; they soon wilt and die. The toxin may remain in the soil after the walnut roots are dead.

APPENDIX
SEED CATALOGUES

Almost all of the tomato varieties described in this book are sold through one or more of the following mail-order catalogues. Code numbers at the end of each variety description refer to numbers 1 through 40 below.

UNITED STATES

1. Agway, Inc.
   RD4, Zeager Road
   Elizabethtown, Pennsylvania
   17022

2. Archias Seed Store Corporation
   P.O. Box 109
   Sedalia, Missouri 65301

3. Burgess Seed and Plant Company
   905 Four Seasons Road
   Bloomington, Illinois 61701

4. W. Atlee Burpee Company
   Warminster, Pennsylvania 18974

5. D. V. Burrell Seed Growers
   Company
   Rocky Ford, Colorado 81067

6. Comstock, Ferre & Company
   263 Main Street
   Wethersfield, Connecticut 06109

7. DeGiorgi Company, Inc.
   Council Bluffs, Iowa 51501

8. Epicure Seeds Ltd.
   P.O. Box 450
   Brewster, New York 10509

9. Farmer Seed & Nursery
   Fairbault, Minnesota 55021

10. Henry Field Seed & Nursery
    Company
    Shenandoah, Iowa 51602

11. Grace's Gardens
    10 Bay St., Dept. 130
    Westport, Connecticut 06880

12. Gurney's Seed & Nursery
    Company
    Yankton, South Dakota 57079

13. Joseph Harris Company, Inc.
    3670 Buffalo Road
    Moreton Farm
    Rochester, New York 14624

14. The Charles C. Hart Seed
    Company
    P.O. Box 9169
    Wethersfield, Connecticut 06109

15. Hastings, Seedsman to the South
    434 Marietta Street, N.W.
    P.O. Box 4274
    Atlanta, Georgia 30302

16. J. L. Hudson, Seedsman
    P.O. Box 1058
    Redwood City, California 94064

17. Jackson & Perkins
    Medford, Oregon 97501

18. Johnny's Selected Seeds
    Albion, Maine 04910

19. J. W. Jung Seed Company
    Randolph, Wisconsin 53956

193

20. Lockhart Seeds
P.O. Box 1361
Stockton, California 95205

21. Earl May Seed & Nursery
Company
Shenandoah, Iowa 51603

22. The Meyer Seed Company
600 South Caroline Street
Baltimore, Maryland 21231

23. Mountain Valley Seeds & Nursery
Company
2015 North Main
North Logan, Utah 84321

24. Nichols Garden Nursery
1190 North Pacific Highway
Albany, Oregon 97321

25. L. L. Olds Seed Company
P.O. Box 7790
2901 Packers Avenue
Madison, Wisconsin 53707

26. George W. Park Seed
Company, Inc.
P.O. Box 31
Greenwood, South Carolina 29649

27. Plants of the Southwest
1570 Pacheco Street
Santa Fe, New Mexico 87501

28. Porter & Son, Seedsmen
P.O. Box 104
Stephenville, Texas 76401

29. The Rocky Mountain Seed
Company
1325–15th Street
Denver, Colorado 80217

30. Seedway, Inc.
Hall, New York 14463

31. R. H. Shumway, Seedsman, Inc.
P.O. Box 777
Rockford, Illinois 61105

32. Stokes Seeds, Inc.
737 Main Street
P.O. Box 548
Buffalo, New York 14240

33. George Tait & Sons, Inc.
900 Tidewater Drive
Norfolk, Virginia 23504

34. Otis Twilley Seed Company, Inc.
P.O. Box 65
Trevose, Pennsylvania 19047

35. The Urban Farmer, Inc.
22000 Halburton Road
Beachwood, Ohio 44122

CANADA

36. Alberta Nurseries & Seeds Ltd.
P.O. Box 20
Bowden, Alberta
Canada T0M 0K0

37. McFayden Seeds
P.O. Box 1800
Brandon, Manitoba
Canada R7A 6A4

38. W. H. Perron & Company, Ltd.
515 Boulevard Labelle
Laval, Quebec
Canada H7V 2T3

39. T&T Seeds, Ltd.
120 Lombard Avenue
Winnipeg, Manitoba
Canada R3B 0W3

40. Vesey's Seeds Ltd.
York, Prince Edward Island
Canada C0A IP0

INDEX

ABOUT THE AUTHOR

Fred DuBose, a native Texan, is a writer and book editor who has lived and grown tomatoes in Connecticut, Europe, Australia, and the South Pacific island kingdom of Tonga. He currently resides in Atlanta, Georgia, where he gardens with the help of his two daughters, Kate (12) and Polly (7).